Seabirds of Hawaii

Seabirds of Hawaii

Natural History and Conservation

CRAIG S. HARRISON

Comstock Publishing Associates A DIVISION OF

Cornell University Press ITHACA AND LONDON

To Mom and Dad

THIS BOOK IS PUBLISHED WITH THE AID OF GRANTS FROM THE HAROLD K. L. CASTLE FOUNDATION AND THE COOKE FOUNDATION, LTD.

First published 1990 by Cornell University Press.

Printed in the United States of America.

Color plates printed in Hong Kong.

Library of Congress Cataloging-in-Publication Data

Harrison, Craig S.
Seabirds of Hawaii : natural history and conservation / Craig S. Harrison
p. cm.
Includes bibliographical references.
ISBN 0-8014-2449-6 (alk. paper)
1. Sea birds—Hawaii. 2. Birds, Protection of—Hawaii.
I. Title.
QL684.H3H37 1990
598.29'24'09969—dc20 89-71222

∞The paper used in this publication meets the minimum requirements of the American National Standard for Permanence of Paper for Printed Library Materials Z39.48–1984.

CONTENTS

Color plates follow p. 142

Preface ix

Part I The Environment and Humans

1 *The Islands* 3

2 *The Sea* 19

3 *The Humans* 29

Part II Comparative Biology of Hawaiian Seabirds

4 *Origin and Adaptations of Hawaiian Seabirds* 43

5 *Populations* 53

6 *Breeding Ecology* 67

7 *Feeding Ecology* 83

8 *Pelagic Ecology: Life at Sea* 98

Part III Hawaiian Seabirds: Family Groups and Species

9 *Albatrosses: Family Diomedeidae* 107

10 *Shearwaters and Gadfly Petrels: Family Procellariidae* 120

11 *Storm-Petrels: Family Oceanitidae* 135
12 *Frigatebirds: Family Fregatidae* 143
13 *Boobies: Family Sulidae* 153
14 *Tropicbirds: Family Phaethontidae* 166
15 *Terns and Noddies: Family Laridae (Subfamily Sterninae)* 175

Part IV Conservation

16 *Conservation on the Islands* 191
17 *Conservation at Sea* 207
18 *Conservation Dilemmas* 219

APPENDIX: *Some Common and Scientific Names* 229

Selected Bibliography 233
Index 239

FIGURES AND TABLES

FIGURES

1. The Hawaiian archipelago *5*
2. The North Pacific *7*
3. The island of Hawaii *10*
4. Maui *11*
5. Molokai *11*
6. Lanai *12*
7. Oahu *12*
8. Kauai *13*
9. Bills of some Hawaiian seabirds *51*
10. Nesting habitats of Hawaiian seabirds *71*
11. Feeding methods of Hawaiian seabirds *86*
12. Diets of Hawaiian seabirds *88*
13. Proportions of fishes consumed by Hawaiian seabirds *90*
14. Estimated annual consumption of marine resources by Hawaiian seabirds *96*
15. July ranges of adult Laysan albatrosses and black-footed albatrosses *111*
16. The exclusive economic zone around Hawaii *212*

TABLES

1. Estimated numbers of breeding pairs, Northwestern Hawaiian Islands *56*
2. Estimated numbers of breeding pairs, Kauai, Niihau, and vicinity *57*
3. Estimated numbers of breeding pairs, Oahu and vicinity *58*
4. Estimated numbers of breeding pairs, Maui and vicinity *59*
5. Estimated numbers of breeding pairs, Kahoolawe and vicinity *60*
6. Estimated numbers of breeding pairs, Molokai and vicinity *60*
7. Estimated numbers of breeding pairs, Lanai and vicinity *60*
8. Estimated numbers of breeding pairs, Hawaii and vicinity *61*

PREFACE

Much of this book stems from the years I spent studying the wildlife of the Northwestern Hawaiian Islands as an employee of the U.S. Fish and Wildlife Service. After the focus of my life shifted to Oahu, I was pleasantly surprised to find seabirds in unsuspected places. From the vantage of my fifteenth-floor law office on Merchant Street I could see brown boobies plunging into Honolulu Harbor and great frigatebirds soaring above Sand Island. White terns fluttered in the trees around the state capitol, and once a pair of startled white-tailed tropicbirds winged their way *mauka* (toward the mountains) up Fort Street Mall. At home in Kaneohe I could watch red-footed boobies and frigates above Kaneohe Bay, and if I listened carefully on still summer nights I could sometimes hear the plaintive cries of sooty terns returning to Moku Manu. I hope this book conveys some of the wonder of Hawaii's seabirds and will help to stimulate improvements in the ways humans manage wildlife in Hawaii.

Many friends and colleagues contributed to my interest in the wildlife of Hawaii over the years. My collaboration with Richard S. Shomura, Thomas S. Hida, and Michael P. Seki of the National Marine Fisheries Service helped me to understand the feeding ecology of seabirds. The camaraderie among Maura B. Naughton, Audrey Newman, Eric P. Knudtson, and Mark J. Rauzon rendered the bureaucracy of the U.S. Fish and Wildlife Service endurable during years of fieldwork on remote islands. Roger B. Clapp, the late Ralph W. Schreiber, G. Causey Whittow, and Alan Ziegler stimulated thought-provoking discussions of arcane aspects of tropical seabird biology. David H. Woodside, a man of the *'aina*, taught me much about the natural history of Hawaii in the field. David C. Callies, John P. Craven, Judge Michael Town, and Jon M. Van Dyke of the

University of Hawaii Law School refined my understanding of environmental law and of the law of the sea.

Cameron B. Kepler generously allowed use of his unpublished data on the seabird populations of Maui and Molokai and, along with John Clark, Daniel Moriarty, and Thomas C. Telfer, improved the population estimates for the main islands. Thane Pratt and Ronald Walker gave me access to the seabird records of the Department of Land and Natural Resources. Kortni Buck drew the pie diagrams and some of the maps. The National Marine Fisheries Service provided the map that depicts the exclusive economic zone, and the U.S. Fish and Wildlife Service provided Figures 10 and 11. I am grateful to the Wildlife Society for permission to reproduce Figure 11. Mark J. Rauzon drew most of the line drawings of seabirds. Marina Chang tracked down several rare documents in the Library of Congress. It is a pleasure to thank all of these people for their assistance.

C. S. H.

Kaneohe, Hawaii
Arlington, Virginia

White-tailed tropicbird along the Na Pali coast

PART I

The Environment and Humans

On landing, we were much annoyed by the birds, many of which made their attack flying, while others ran after us, pecking at our legs: it was with great difficulty we could keep them off, even with our canes. . . . The heat of the day was excessive, and, almost at every step, we sunk up to our knees in holes, that were concealed by overgrown creeping plants, and contained nests, as we supposed, of various birds; for we often heard their cries under our feet from being trampled upon. . . . [Lisianksi] promises nothing to the adventurous voyager but certain danger in the first instance, and almost unavoidable destruction. . . .

—Captain Urey Lisianski, October 10, 1805

1 THE ISLANDS

The National Oceanographic and Atmospheric Administration's ship *Townsend Cromwell* rolled and pitched in the choppy late-February seas as squalls alternated with clear patches in the turquoise-charcoal expanse of sea and sky. On the lee side of the bridge I was protected from the sloppy weather while awaiting my first glimpse of the famous Laysan Island. It was unimpressive. Except for one grove of palms, the low, flat band that appeared on the horizon increased little in relief as we carefully approached through the coral shallows. Dozens of Laysan and black-footed albatrosses circled the research vessel while we waited for the seas to moderate so that the crew could safely put a zodiac over the side. With the wind and swells bearing from the northwest, the eastern shore provided sufficient protection to allow our team of biologists to land. The rough days at sea were worth the trouble. This flat, sandy island greeted me with one of the finest remaining wildlife spectacles on earth—hundreds of thousands of clacking albatrosses, subterranean colonies of wedge-tailed shearwaters that moaned eerily at night, bevies of inquisitive white terns, scores of endangered monk seals dozing on the beach. Soon enough I also met clouds of pesky house- and blowflies that seemed to be the size of bats when they swarmed around my face.

Although the Hawaiian archipelago stretches almost 2,500 kilometers across the North Pacific Ocean, most people think of it as consisting of the eight main or high islands in the southeast portion of the archipelago, which includes over 99.9 percent of its 16,600 square kilometers of land mass and virtually all of the human population. But Laysan exemplifies another Hawaii—Hawaii's seabird colonies. That Hawaii consists of tiny islets and stacks offshore the main islands, such as Molokini and Moku Manu. And it includes emergent rocks, low-lying coral reefs, and sand spits northwest of Kauai with romantic-sounding names such as French Frigate Shoals, Gardner Pinnacles, and Pearl and Hermes Reef.

No other volcanic islands are so far from a continent as the Hawaiian Islands. They are not even close to other islands of appreciable size. Virtually alone in the North Pacific, the Hawaiian Islands are almost 4,000 kilometers from North America and the Marquesas and over 6,000 kilometers from Japan. The nearest neighbors are small islands—Johnston, Wake, Marcus (Minami Torishima), Palmyra. Geological evidence indicates that the Hawaiian chain was never a continuous strip of land, nor has it ever been connected to any continent. No "vanished" islands can reasonably be imagined between Hawaii and North America. The Hawaiian archipelago comprises about 132 islands, reefs, and shoals that straddle the Tropic of Cancer. Of these, twenty-seven or so, those from Nihoa to Kure Atoll, are designated the Northwestern Hawaiian Islands (Figure 1) by the geographer of the State of Hawaii. The precise number changes when sand spits form, disappear, or merge after severe winter storms. Disappearing Island in French Frigate Shoals is appropriately named, and Seal-Kittery in Pearl and Hermes Reef was once two distinct islands.

The Volcanic Origin of the Hawaiian Islands

The Hawaiian Islands are almost entirely volcanic in origin—sedimentary rocks form only a narrow ridge around the edges of some large islands. The vast majority of the volcanic rocks are lava flows, formed by outpourings of liquid magma. A few were formed of fragments thrown out by volcanic explosions. The volcanic nature of the islands is inescapable to any resident or visitor. Kilauea Volcano, on the island of Hawaii, is the most continuously active volcano on earth, and such volcanic features as cinder and tuff cones, domes, and giant shields are prominent on most main islands. Tremors are common and can be felt daily on the island of Hawaii.

Oceanic islands are almost always volcanic in origin. What forces control the formation of islands that must rise 5,000 meters from the ocean floor merely to break the sea surface? The theory of plate tectonics, formulated by earth scientists in the late 1960s, provides simple answers to such questions. Plate tectonics unifies observations concerning continental drift and sea-floor spreading and provides an explanation of the evolution of ocean basins and continents over the past 200 million years. The theory of plate tectonics is as important a unifying theory within the earth sciences as the theory of evolution is to the biological sciences.

Earth scientists now recognize that the outer layer of the earth's surface is divided into many blocks or plates. Most plates consist of continental land masses and adjacent ocean floor, but some consist entirely of oceanic areas. Most of the Pacific basin, including the Hawaiian archipelago, is located on the

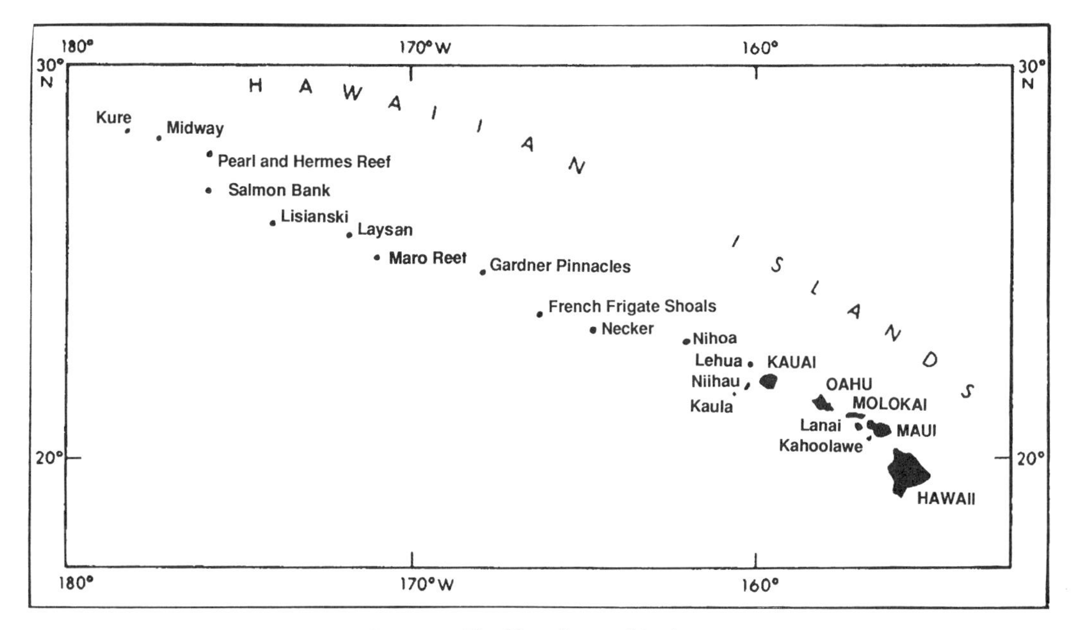

Figure 1. The Hawaiian archipelago

Pacific plate, the largest on earth. Like all plates, the Pacific plate is rigid and moves over a layer of molten earth some 60 kilometers below the surface. Where two plates converge, as they do near the Marianas Islands, a trench is created where one plate overrides another. Where two plates diverge, partially molten material rises to the surface through the weakened crust and forms ridges.

The unique and crucial phenomenon that has created the Hawaiian archipelago is the existence of a hot spot in the Pacific plate, which is currently located in the vicinity of the island of Hawaii. A hot spot results from a plume of magma that rises to the earth's surface from deep in the mantle. Geologists recognize sixteen such hot spots on the face of the earth. The molten rock at a hot spot is forced up through a rupture or fault in the basalt rock of the ocean floor. Over hundreds of thousands or millions of years, small amounts of lava ooze out, each new layer piling up on previous layers. Such processes created the Hawaiian islands, some of which rise ten thousand meters from the ocean floor and only half of which are above sea level.

The volcanos of the Hawaiian archipelago were built in a northwest–southeast direction across the North Pacific Ocean floor. The islands are so closely grouped in the chain and have extruded so much lava that they are connected by huge submarine ridges. The ages of basaltic lava on different islands can be estimated by radioisotope methods. The oldest, Kure Atoll, was formed about 30 million years ago. While the hot spot's location has remained fixed, the Pacific plate has gradually moved northwest, shifting the locations of eruptions on the plate until now the only active volcanos are those in the southeastern portion of the chain. Salmon Bank, a submerged volcano near Pearl and Hermes Reef, is estimated to have formed about 20 million years ago. French Frigate Shoals is about 10 million years old.

A similar pattern holds even among the main or high islands. Kauai, at the northwest end, is about 4 million years old; Hawaii is less than a million. Loihi Seamount, some twenty kilometers southeast of Hawaii, is potentially the next Hawaiian island. It is actively building now, and if it increases only about a thousand or so additional meters, it will emerge as an island. This process may occur quite quickly by geological standards, but rather slowly in human terms—estimates vary between 2,000 and 20,000 years. Several of the main islands were formed by a wedding of two separate volcanos, such as the Waianae Mountains and Koolau Mountains on Oahu. Even there the west–east pattern is remarkably constant: the western Waianae range is older than the eastern Koolaus.

Abundant geophysical evidence has established that the Hawaiian chain is related geologically to the Emperor Seamounts, a series of submarine peaks that extend north beyond Kure Atoll as far as the western Aleutians (Figure 2). The Emperors are similar in profile to other oceanic islands, but smaller. They rise several kilometers above the ocean floor with circular or elliptical bases and steep slopes and are grouped in a line running from north to south. The oldest reef corals in the Emperors have been recovered from Koko Guyot, which is estimated to have formed 48 million years ago. Meiji Guyot, the northernmost volcano edifice at 53 degrees north latitude, is 70 million years old. The Emperors emerged over the same hot spot as the Hawaiian chain.

The rocks and atolls of the Northwestern Hawaiian Islands, covering less than fourteen square kilometers, are worn to mere vestiges of their ancient estate. What did the chain look like millions of years ago? As now, it was certainly an archipelago of isolated islands far from any continent, with a tropical or subtropical climate similar to the one we know today. The Northwestern Hawaiian Islands were once high volcanos, but just how high and how large is a matter of conjecture. The size would have fluctuated over time with the dynamic processes of island building and the eventual decay and erosion wrought by wind, rain, and sea. Some of the atolls and shallow banks surrounding the islands rival today's high islands in size, and the former masses of the Northwestern Hawaiian Islands were probably once at least the size of Kauai or Oahu. During glacial periods, huge accumulations of ice dramatically changed the levels of the ocean. Although such changes would have occurred over thousands of years, it is certain that at times the sea level was at least 100 meters lower than it is today. If the sea were at such a level now, the atolls at French Frigate Shoals, Pearl and Hermes Reef, Pioneer Bank, and Salmon Bank would be large islands. It is probable that some 17,000 years ago, during the Wisconsinian glacial period, the shallow channels that today separate Maui, Molokai, and Lanai were dry land, so that the three islands formed a single large one; geologists have named it Maui Nui. Penguin Banks, a submerged shelf west

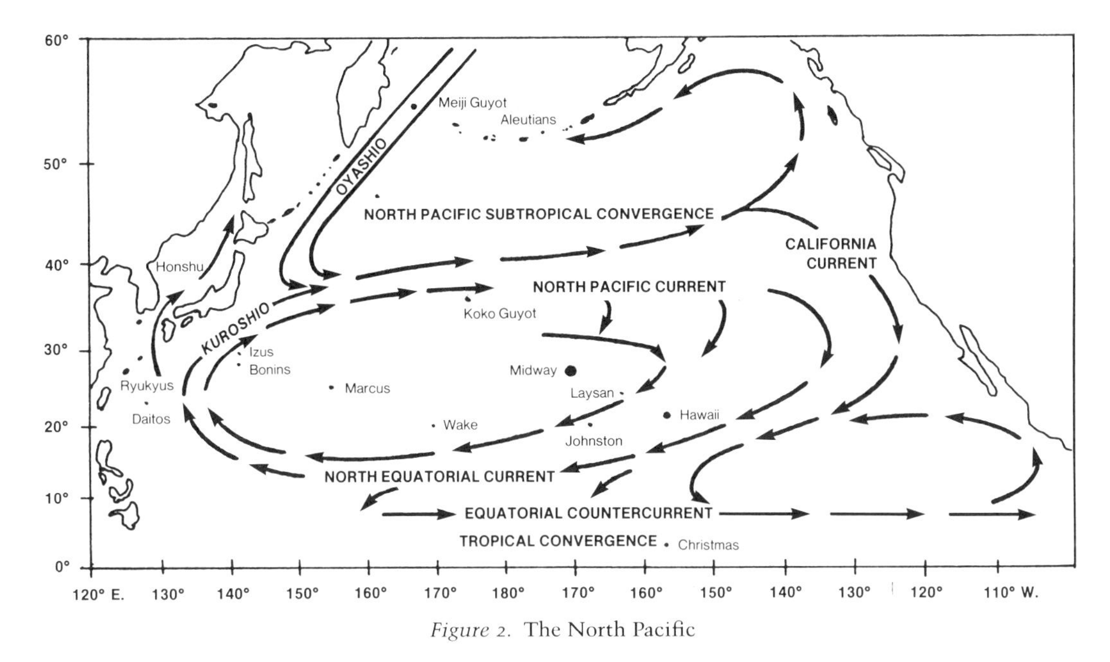

Figure 2. The North Pacific

of Molokai, was then an emerged island. What the sea level can give it can also take away. During warm periods with minimal polar ice caps, sea levels have been at least 30 meters higher than they are today. Such levels, if not accompanied by concomitant growth of coral reefs, would drown all of the present Northwestern Hawaiian islands except outcroppings at Nihoa, Necker, La Perouse Pinnacle, and Gardner Pinnacles. Such levels would also submerge the coastal areas of the main islands, shrinking their land masses substantially. Conceivably, during periods of increased sea levels in the past, Maui and Molokai have each been split into two separate islands.

The transition stage between a high and a low volcanic island passes rapidly in geological terms. Even as periodic ejaculations of molten lava are building a volcanic high island, natural forces begin to chip it away. Wind, rain, and lichens break down the sheets of flat rock to form soil. Waves cut down high sea cliffs, streams carve deep valleys, and after a few million years a once-mighty volcano has begun to disappear beneath the sea. In tropical and subtropical regions such as Hawaii, reef-building corals flourish in the nutritious nearshore waters. As long as the seawater remains above 18 degrees centigrade, corals (relatives of jellyfish) and coralline algae can grow in coastal waters down to depths of 30 meters, depending on the depth that sunlight can penetrate. A reef is a colony of millions of tiny individual animals or polyps, each of which secretes a chalky, cuplike skeleton. As each polyp dies, it leaves behind a calcium carbonate skeleton on which successive generations of corals live. The countless skeletons that are deposited over eons form a calcium carbonate wreath around the drowning island; only the surface layer is alive.

Coral reefs function to ensure the continued existence of most of the Northwestern Hawaiian islands. As long as the rate of coral growth and other sources of limestone production keep pace with the changes in sea level which cause an island to drown, the Northwestern Hawaiian Islands will remain emerged. The depth of coral deposits that cap the basalt bedrock of the Northwestern Hawaiian Islands has not been determined but is probably considerable. Borings at Eniwetok and Bikini atolls in the Central Pacific went through 1,400 meters of coral deposits before hitting bedrock. A threshold for atoll formation, called a Darwin point, exists today near Kure Atoll. North of Kure, growth and development of coral in the cooler waters are too slow to keep pace with the forces that erode and drown islands. Research by Richard Grigg of the Hawaiian Institute of Marine Biology suggests that the Darwin point has remained near its present location at 29 degrees north latitude for 20 million years or so. The principal submerged peaks of the Emperor Seamounts are capped by drowned coral reefs that were unable to grow fast enough once the Pacific plate pushed the reefs into the cold North Pacific waters beyond the Darwin point.

The Main Hawaiian Islands

Little is known about the numbers or even species of seabirds that nested on the main Hawaiian islands when Westerners arrived. Our knowledge about the situation before the archipelago was colonized by the Polynesians is restricted to what we can learn from fossils and middens. Studies by Storrs L. Olson of the Smithsonian Institution have greatly expanded our knowledge of the fossil avifauna in the main islands. Today most of the main islands' nesting seabirds are found in relatively high mountain areas. Cliff-facing ledges and crater walls such as those on Haleakala Crater on Maui, at Mauna Loa and Mauna Kea on Hawaii, and in the canyons of Hanapepe Valley on Kauai sometimes harbor nesting birds, as do predator-free coastal headlands such as Kaholo Pali, Lanai, and Crater Hill, Kauai. Such areas are usually sparsely vegetated, with only a few grasses and ferns interspersed among the rocks. Some coastal areas support colonies of red-footed boobies, which nest on shrubs well above ground level. Burrowing birds require nest sites free from predators, and such habitat is now particularly rare in the coastal portions of the main islands. The vegetation of the main islands is generally varied, and much of the Hawaiian landscape has been usurped by alien species, especially in coastal areas.

The primary factor that determines the locations of seabird colonies on the main islands is the absence of introduced predators. In their pristine prehuman condition, the Hawaiian Islands were home to a single terrestrial mammal, the hoary bat. No land-based predator posed a threat to nesting seabirds. Ancient Polynesians brought dogs, pigs, and probably Polynesian rats to the islands. Much later, Westerners brought European pigs, black rats, Norway rats, cats, mongooses, and mosquitos, which harbor avian malaria in their salivary glands. Many alien mammals eat eggs, chicks, and adult birds. High mountain rocky outcrops and canyon walls provide nesting sites free from such predators. Similarly, dense stands of uluhe ferns on Kauai provide protection for Newell's shearwaters at elevations as low as 150 meters. Elsewhere, small colonies may form where local topography affords protection. White terns in recent years have nested in various exotic trees in the parks of urban Honolulu, especially Kapiolani Park in Waikiki. Red-footed boobies nest in shrubs at Ulupa'u Head in the Kaneohe Marine Corps Air Station, Oahu, and along the steep cliff face at Kilauea Point, Kauai. Wedge-tailed shearwaters nest at many coastal sites on mongoose-free Kauai. They fare best on a flat-topped peninsula at Kilauea Point, where refuge managers use fences and traps to exclude dogs and cats. Cliff faces in isolated headlands such as Waihe'e Point, Maui, also provide sufficient natural protection for wedge-tails to nest, and undoubtedly biologists will discover other small colonies when the coasts of the main islands, especially Niihau, have been carefully surveyed. Several caves and cliffs along Kauai's Na Pali coast provide sufficient protection for small colonies of black noddies.

Suitable protection tends to be found on the small islets that lie just offshore the main islands. The tuff cones and sea stacks are rarely large enough to support populations of medium-sized predators such as dogs and cats, but rats are frequently a problem. Several small colonies of seabirds are found offshore Hawaii (Figure 3), Maui (Figure 4), Molokai (Figure 5), and Lanai (Figure 6). The difficulty of landing on such islets makes information scarce but also provides the protection that is needed for successful nesting. Thirteen colonies, two of them large and important, are located within a kilometer of windward Oahu (Figure 7). Manana (Rabbit) Island just offshore Waimanalo consists of two tuff cones, one of which encompasses about 63 acres. Manana is vegetated primarily with grasses, but wild tobacco and several palm trees also grow there. European rabbits were introduced by 1900 and probably have restricted plants to such an extent that they no longer attract shrub-nesting seabirds. Rabbits and mice apparently have not posed serious problems to the large colony of five ground- and burrow-nesting seabird species there, and rabbits may have died out naturally during the 1980s. As Manana can easily be reached by small boat from Waimanalo pier, it is a convenient location for scientific studies and naturalist field trips. Moku Manu (Bird Island), an eroded volcanic cone just offshore Ulupa'u Head, Kaneohe, consists of an eighteen-acre main island adjacent to a three-acre rock. It has sufficient shrubbery to allow red-footed boobies to nest, in addition to at least eleven other species. Landing is problematic at best, and I bear scars from clambering ashore there during a biological survey in heavy surf. Mokoli'i (Chinaman's Hat), Moku'auia (Goat Island), and Popoi'a (Flat Island) harbor rats, possibly because of their proximity to shore.

There are several seabird colonies on Kauai (Figure 8) and Niihau. Lehua and Niihau (Figure 1) were once a single island, but when the land that connected them was worn down to an underwater shelf, Lehua was created as a 291-acre separate island. Despite the presence of European rabbits and rats, Lehua has a colony of at least ten seabird species. Kaula Island, a 136-acre rock about 40

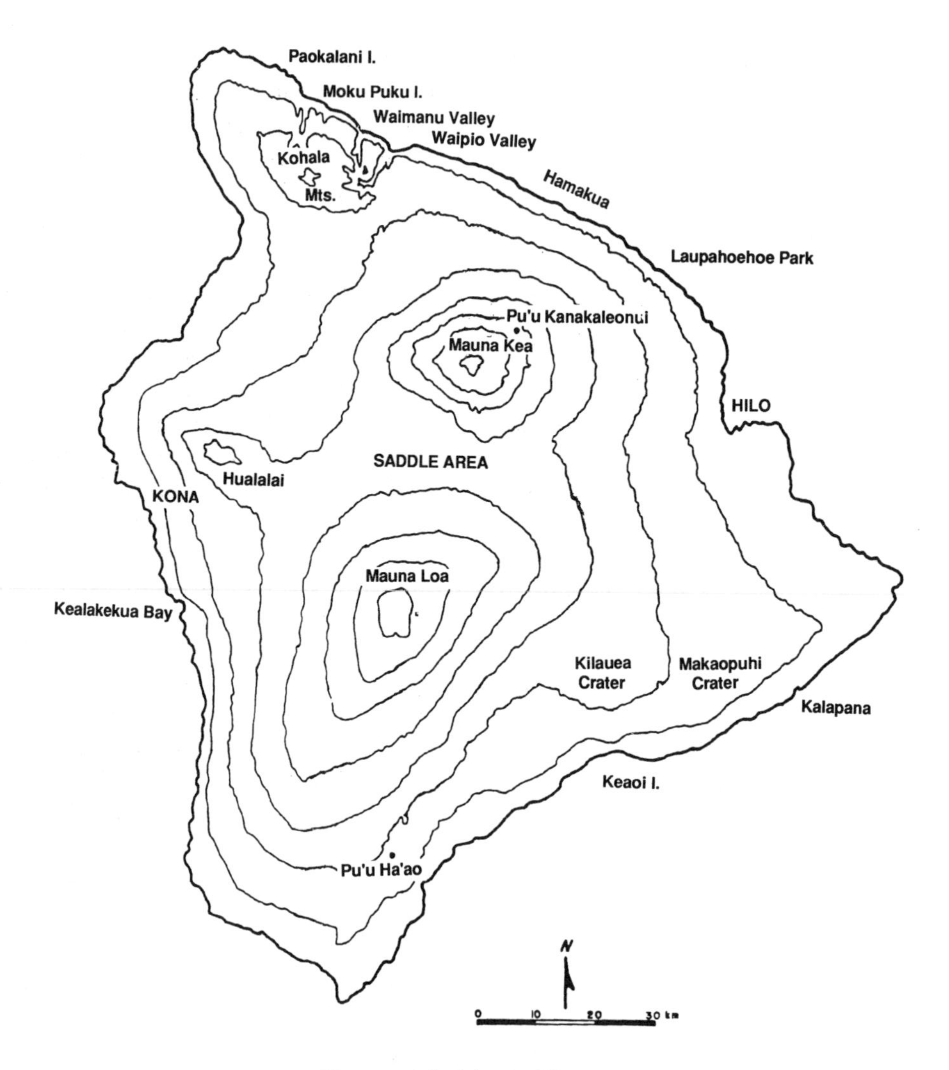

Figure 3. The island of Hawaii

kilometers southwest of Niihau, is also an important seabird colony. Although it has been used by U.S. Navy and Marine Corps aircraft for bombing and strafing practice since 1952, at least sixteen species of seabirds breed there. An unknown species of rat was present there in 1938.

The Northwestern Hawaiian Islands

The Northwestern Hawaiian Islands are quite small. The plant community consists of the typical beach species that are distributed throughout the tropics by coastal currents. These islands, however, are home to several endemic plants

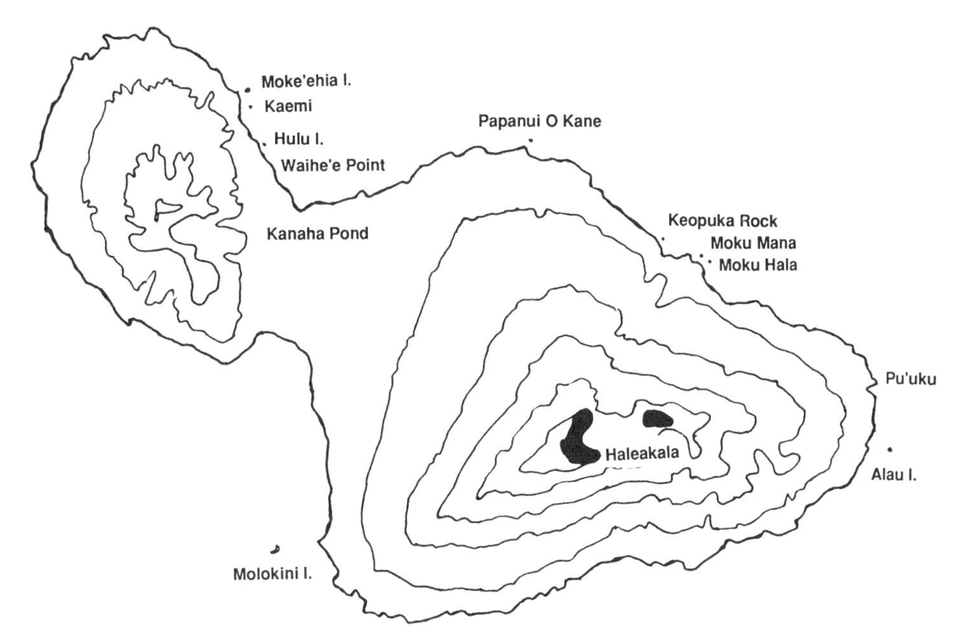

Figure 4. Maui. Shading represents nesting areas of dark-rumped petrels.

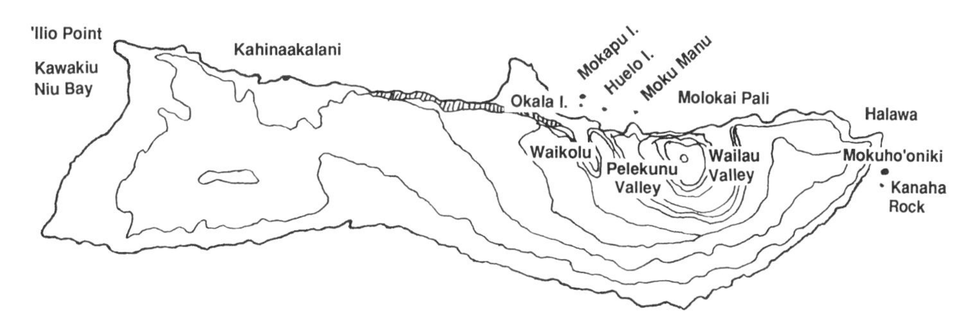

Figure 5. Molokai

and birds and a Hawaiian monk seal that is found nowhere else, as well as to 14 million of Hawaii's 15 million seabirds, including eighteen of Hawaii's twenty-two species.

Nihoa Island

Nihoa Island, the easternmost of the Northwestern Hawaiian islands, is 400 kilometers from Honolulu. Landings on Nihoa can be difficult, especially during winter, and consequently much of its natural history is poorly known. It sits at the southwest edge of a submarine bank where water depths range from 50 to

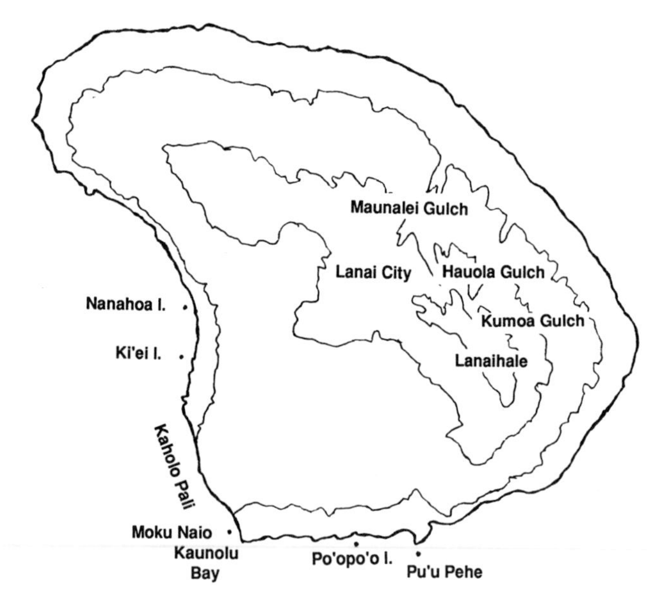

Figure 6. Lanai

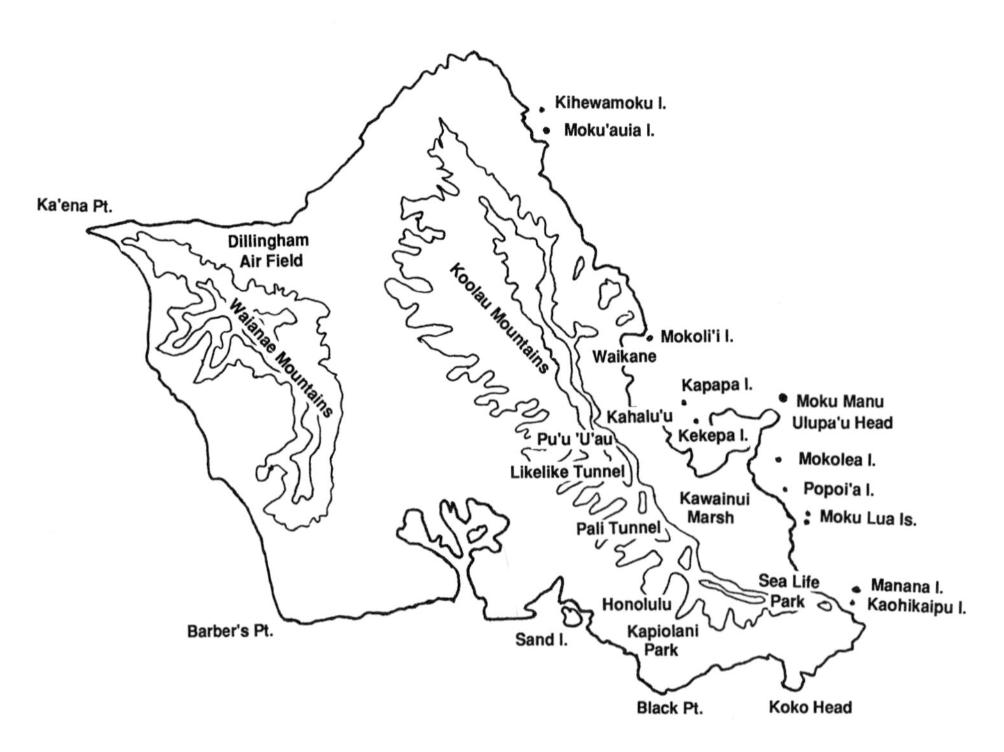

Figure 7. Oahu

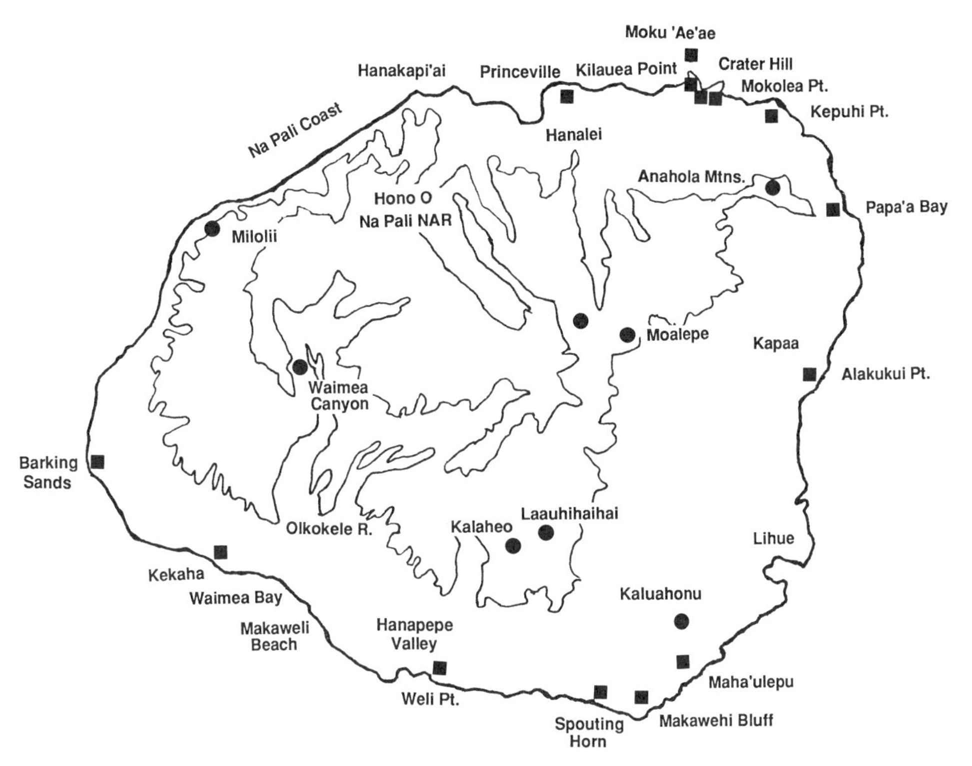

Figure 8. Kauai. Newell's shearwater colonies are represented by circles, wedge-tailed shearwater colonies by squares.

70 meters. With an area of 156 acres, Nihoa is the largest of the lava islands northwest of Niihau. This precipitous remnant of a volcanic cone has steep slopes and rocky outcroppings. Nihoa is too sheer to have a fringing reef. The northern and eastern edges are vertical cliffs, the result of waves driven into the island by the prevailing trade winds. Fanning out from Adam's Bay are six well-developed valleys that are densely vegetated with low shrubs such as goosefoot, nightshade, and ilima and an endemic loulou fan palm. Bunchgrasses are common on the ridge areas but Miller's Peak (300 meters in elevation) and Tanager Peak are mostly bare. As the island has no streams and only a few seeps, it is subject to drought between storms. Nihoa supports a remarkable flora and fauna that occur nowhere else, including four plants, 35 arthropods, about 3,000 Nihoa finches (a honeycreeper), and 350 Nihoa millerbirds. Populations of the endemic birds seem to fluctuate widely. Small numbers of Hawaiian monk seals and green sea turtles bask on Nihoa's small beaches but rarely reproduce there.

Necker Island

Necker Island is a sharply rising ridge of volcanic rock which, like all of the other Northwestern Hawaiian islands, is a remnant of a volcanic cone. This rocky island lacks a fringing reef and is located on the northwestern end of a large, shallow bank some 120 kilometers southeast of French Frigate Shoals. With an area of only 41 acres, it is one-quarter the size of Nihoa. Necker has steep slopes, shallow valleys, and a craggy, wave-sculpted appearance. The island consists of two parts. The principal one is a 1,300-meter-long ridge that varies in width from 60 to 200 meters. The lesser part, Northwest Cape, is separated from the main island by a gap that is barely above sea level. Landings at the gap can be dangerous during high surf. Necker has no trees and its thin soil has permitted the establishment of only five plant species, two of which (goosefoot and ohai) are shrubs. No land birds are found there, probably because the island is too small to support them. Green sea turtles bask on the gap area of Necker, but they cannot nest on the island because it has no sandy beaches. Increasing numbers of Hawaiian monk seals have hauled themselves out on Necker during recent years, and some pups are born there.

French Frigate Shoals

French Frigate Shoals is located at the midpoint of the Hawaiian archipelago. The atoll is unusual in that La Perouse Pinnacle, a one-acre rock standing 40 meters above sea level, still stands as the last vestige of the high island from which the atoll was derived. French Frigate Shoals is an imperfect, crescent-shaped atoll on an oval platform 40 meters deep. The crescentic reef is double. The outer arc, 56 kilometers long, is almost continuous, whereas the smaller inner arc is broken. In addition to La Perouse, twelve small, sandy islets exist in the atoll today. Topographical changes have taken place periodically during the past century. Gin, Little Gin, and Disappearing islands are inundated from time to time, and Bare, Mullet, Near, Round, and Shark islands are essentially continuously shifting sandbars. Whale-Skate Island consisted of two separate islands as recently as 1923 but they are now "permanently" joined into a single islet, at least until a major storm rearranges the atoll.

The transitory nature of many of the islets in French Frigate Shoals is reflected in the vegetation. Only about a third of the 111 acres of emergent land within the atoll is vegetated, and well-established vegetation is restricted to Tern, Trig, East, and Whale-Skate. Ironwood trees have been planted along the runway at Tern and a few beach magnolia and beach heliotrope shrubs grow on some of the islets. Otherwise, all of the vegetation consists of typical low herbaceous atoll plants, such as morning-glory, puncture vine, purslane, and *Boerhavia.* Tern Island originally covered about 11 acres but was converted into a 57-acre spoil island to serve as an airfield during World War II. Much of its surface today is covered by the airstrip, fuel tanks, and buildings. There are no native land birds within the atoll, but the islands are crucial to the continued existence of Hawaiian monk seals and green sea turtles. William Gilmartin estimated a population of about 800 monk seals there during the late 1980s, by far the largest population anywhere. Each year 300 or so green sea turtles nest within French Frigate Shoals and account for about 90 percent of the sea turtles nesting in the Hawaiian Islands.

Gardner Pinnacles

Gardner Pinnacles is the smallest island group in the Northwestern Hawaiian Islands, comprising two volcanic rocks totaling three acres. It is the westernmost volcanic island in the Hawaiian archipelago and almost certainly the oldest piece of emerged lava in the chain. Gardner Pinnacles lack a fringing reef and are located on the northeast portion of a large submerged bank with depths ranging between 18 and 80 meters. The bank is roughly 32 by 80 kilometers in extent and would be larger than Maui if it were entirely emerged. The larger basalt pinnacle is 58 meters high and from a distance appears to be capped with snow, which on closer inspection is revealed to be liberal coats of guano. The rocks are largely barren, and only a few plants such as purslane and ohai have successfully gained a roothold. Access to the islands is extremely difficult, and scientists aboard research vessels rarely even attempt to land. A few monk seals and green sea turtles can be seen nearby, but they do not pup or nest there.

Maro Reef

Maro Reef, between Gardner Pinnacles and Laysan (Figure 1), is an oval-shaped coral bank 45 by 29 kilometers. The entire area consists of submerged reefs and coral heads except for a single rock that barely protrudes above sea level. Among the submerged banks of the Northwestern Hawaiian Islands, Maro Reef is unique in the extensiveness of its shallow portions.

Laysan Island

Laysan Island is located about 1,100 kilometers northwest of Honolulu. Like other islands in the chain, it sits on the flattened top of what was once a massive volcanic peak formed perhaps in the Miocene. Shallow water less than ten meters deep extends in all directions for eight kilometers, and the bank defined by the 200-meter contour line around the island encloses an area of over 500 square kilometers. Laysan is somewhat unusual in being slightly elevated, so that it has a larger land mass than most other atolls. What would normally be a lagoon inside a barrier reef is primarily land, some of which is ten meters above sea level. A fringing reef surrounds the island, varying from 100 to 500 meters in width at the northwest end. With an area of 913 acres, it is the largest of the Northwestern Hawaiian islands. About one-fifth of the area is a lagoon in a central depression which harbors brine shrimp and is much saltier than the ocean.

Laysan is roughly rectangular and is ringed by large, stabilized sand dunes. Except for the coastal dunes, it is generally well vegetated. The beaches have patches of beach magnolia and beach morning-glory. Clumps of tall bunchgrass characterize the area that slopes gently toward the lagoon. Low-lying succulents such as carpetweed grow in the mudflats and along the edge of the lagoon. The only trees on Laysan are a grove of introduced coconut trees in the northwest corner of the island and a single stunted ironwood. The relatively large area combined with its isolation have permitted Laysan to develop unique flora and fauna. Five endemic plants and an equal number of endemic birds occurred before the turn of the century, but various depredations brought the extinction of Laysan sandalwood, loulou fan palms, rails, millerbirds, and honeyeaters. Today about 500 endemic Laysan ducks and 10,000 endemic Laysan finches live there. The island is an important rookery for Hawaiian monk seals and a minor nesting area for green sea turtles. Unfortunately, humans have introduced house- and blowflies, which are noxious and constant irritants to visitors.

Lisianski Island

Lisianski Island sits on the northern edge of Neva Shoal, a 170-square-kilometer reef bank. Another major submerged reef, Pioneer Bank, lies 35 kilometers to the east. Like Laysan, Lisianski is an upraised atoll with a central depression, but unlike Laysan, it has lacked a central lagoon, at least during historic times. Lisianski is a flat sand-and-coral island. With an area of 450 acres, it is larger than any of the Northwestern Hawaiian islands except Laysan and either Midway island. Few of this oval island's physical features are noteworthy. The eastern beach has an exposed ledge of reef rock, but otherwise the island is fringed by a narrow sandy beach. The only trees are a few scattered ironwoods, several of which are dead. A rim of beach magnolia grows along its entire perimeter, the most dense growth being found in the sand dunes in the northeast portion. The interior is covered with a lush growth of bunchgrass and several associated plants. Beach morning-glory and spawling mats of puncture vines are also common. Lisianski has no endemic plants or birds, possibly because of widespread destruction of vegetation during the early years of the twentieth century. As on Laysan, humans are plagued by swarms of alien flies. Lisianski is an important rookery for Hawaiian monk seals and harbors a few green sea turtle nests.

Pearl and Hermes Reef

Pearl and Hermes Reef is a large classical atoll in the center of a submerged bank located almost 2,000 kilometers northwest of Honolulu and 160 kilometers southeast of Midway Islands. The fringing reef encompasses a shallow lagoon 70 kilometers in circumference which opens to the west. Like French Frigate Shoals, Pearl and Hermes Reef has undergone considerable changes in topography over time. Although an 1858 map of the atoll shows twelve islands, only eight are present today. As recently as the 1960s a ninth island existed, but Seal and Kittery islands have since merged. The ephemeral nature of the 85 acres of emerged land is exemplified by Bird, Planetree, and Sand islands, each of which is a barren, continuously shifting sandbar. The other islands are sufficiently permanent to permit the growth of low foliage such as bunchgrass, purslane, nightshade, puncture vine, and carpetweed. There are no trees today on the atoll and visiting biologists are blinded by the noonday glare off the flat expanse of coral rubble. A few scrawny beach magnolia and nightshade plants grow to sufficient proportions to be deemed shrubs if one adopts a charitable definition. Refuge managers introduced Laysan finches to Pearl and Hermes Reef in 1967 and they still survive on Southeast and North, the two largest islets. The Hawaiian monk seal population has declined precipitously since the 1960s in the northwestern portion of the Hawaiian archipelago but a few seals continue to live at Pearl and Hermes. Green sea turtles are common and several nest there each year.

Midway Islands

The Midway Islands are the most recognized location in the Northwestern Hawaiian Islands. Situated over 2,000 kilometers from Honolulu and almost precisely halfway between North America and Asia, they have provided strategic and commercial benefits to the United States for almost a century. The atoll is a nearly circular rim of coral reef, about eight kilometers in diameter. Most of the lagoon is an expanse of shallow water, but the central portion deepens to 21 meters. Much of the northeast portion of the reef is fairly wide and stands sufficiently above sea level to be visible from the islands. Sand and Eastern (Brooke's) islands lie close to the southern rim of the atoll, separated by two semipermanent spit islands. The islands cover 5.5 square kilometers; the largest is Sand. Before human contact, Eastern had vegetation typical of low coral atolls, such as bunchgrasses and beach magnolia. George C. Munro and Henry Palmer surveyed Eastern in 1891 and found it to be the greenest of the Northwestern Hawaiian islands. In its natural condition, Sand consisted primarily of huge dunes of white, shifting sand and was well named. Henry Palmer, a member of the Rothschild expedition, described Sand on July 11, 1891:

> Although this island is comparatively large, it is the most desolate place I was ever on. There is hardly any vegetation except for a few tufts of grass on the south end, and in rough weather most of the island is under water.

During the early twentieth century, employees of the Commercial Pacific Cable Company transformed the island. They imported 9,000 tons of soil and planted exotics such as ironwood trees and San Francisco grass. Today Sand's towering ironwood trees along Commander Row and near the dilapidated Pan American Hotel rival in size any such trees in their Australian homeland. Except for beach areas, little of Sand or Eastern has natural cover—airfields, harbors, roads, houses, buildings, and bare asphalt cover much of the naval air station and the horizon is broken by antennas, guy wires, and tall trees. Sand is heavily planted with introduced vegetation, including the grass of a nine-hole golf course. Many alien wildlife species have been introduced. Laysan rails and finches flourished on Midway in the 1930s, but were wiped out by black rats and avian malaria at the end of World War II. Today black rat populations can be enormous, and introduced canaries, mice, and mosquitos thrive. Few Hawaiian monk seals and green sea turtles use Midway Atoll.

Kure Atoll

Kure Atoll, the northern- and westernmost of the emerged Hawaiian islands, is the northernmost coral atoll on earth. Coral grows more slowly here than at any other atoll in the Hawaiian archipelago, yet obviously it grows quickly enough to keep pace with erosion and ensure that the island does not drown. The atoll is nearly circular with an outer reef almost completely enclosing a 46-square-kilometer lagoon. The lagoon is fairly shallow and nowhere more than 14 meters deep. Two islands rise on the southern portion of the atoll. Sand is a bare, sandy islet that often splits into several sandbars. Green is a stable, heavily vegetated island encircled by sandy beaches. A dense forest of beach magnolia, up to two meters high, covers much of Green. A United States Coast Guard LORAN station there has necessitated the building of a 1,200-meter runway and the erection of a 200-meter LORAN tower in the central plain of Green. Most of the vegetation at Kure today has been introduced, including ironwood trees and beach heliotropes. Hawaiian monk seals live and reproduce at Kure Atoll, but their population has been declining for many years. Green sea turtles are also seen but they rarely breed here. An unexplained puzzle is the occurrence of Polynesian rats, which may not have been brought by Western ships. Alexander Wetmore believed that they were distributed from island to island in the Pacific as stowaways in the great sailing canoes of the Polynesians, who located atolls such as Kure by searching the skies for green-tinted clouds. Possibly the rats on Kure are the legacy of an errant canoe that otherwise has been lost to history.

2 THE SEA

During my first five-hour flight from the West Coast to Honolulu I periodically glanced out the window of the 747 as the featureless North Pacific Ocean passed by. I became convinced that oceans are the dominant feature of our blue planet. In an era of jet aircraft with iced drinks, magazines, and feature motion pictures to pass the time, it is easy to forget just how much water separates Hawaii from the continents and how insignificant the islands appear from a distance. Yet it is difficult to be far removed from the sea in Hawaii—the evolution of the land, the creatures, and the humans has been intimately tied to the sea. An understanding of the nature of the tropical and subtropical ocean that surrounds the Hawaiian Islands is fundamental to the natural history and ecology of its seabirds. While the waters that bathe the Hawaiian archipelago extend to depths of four kilometers offshore in the north-central Pacific basin, it is the surface waters that are most important to seabirds.

The distribution and abundance of seabirds in all bodies of water depends on the interaction of physical and biological phenomena, and birds usually appear wherever they can find sufficient food and suitable surface water conditions. Seabird species throughout the world are closely associated with particular water masses, so that it is often possible to predict surface water conditions on the basis of the presence of certain birds and vice versa. Like all marine biological communities, the Hawaiian marine ecosystem is characterized largely by the availability of food, which is affected by sunlight, the temperature and salinity of the water, and the nutrients in it. Hawaiian waters have unique characteristics that ultimately prescribe the food chain that supports seabirds.

Global Oceanography

Marine biologists divide the global marine environment into five broad oceanic zones on the basis of latitude: (1) antarctic, (2) subantarctic, (3) subtropic and tropic, (4) boreal (subarctic), and (5) arctic. Ideally, the surface temperature of each zone would be equivalent to a corresponding vegetation or climactic belt on land. But ocean currents drastically distort the idealized boundaries of many zones: the Humboldt current off the west coast of South America extends the cold waters of the subantarctic zone as far north as the equator.

The temperature and chemistry of marine water masses do not vary continuously from one zone to another. Instead, the major zones consist of discrete water masses that are relatively uniform in their temperatures and chemistries. Fairly abrupt changes occur at the borders of adjacent water masses. In these convergence zones, denser water sinks below the surface, and the associated mixing generates choppy seas. Changes may be dramatic, as in the antarctic convergence, where oceanographers have measured temperature changes of five degrees centigrade over a few hundred meters. The North Pacific subtropical convergence is a more typical boundary, with gradual changes in surface waters which extend many kilometers. The precise boundary of each latitudinal zone varies with seasonal and annual weather conditions. The tropical convergence, about 1,300 kilometers south of Honolulu, and the subtropical convergence, about 2,200 kilometers north, are so far from the Hawaiian Islands that the Hawaiian marine ecosystem remains subtropical throughout the year.

Surface water movements cause well-marked zones of convergence and divergence. The intertropical convergence zone is located near the equator in all oceans where both southeast trade winds and northeast trade winds transport surface water toward the equator. In the Pacific, the zone is located north of the equator near Panama and meanders southeast to about 1,500 kilometers south of the equator near Australia. At the intertropical convergence zone, warm water accumulates and is forced downward. Conversely, water upwells wherever divergence occurs. Such vertical movements of water are important to the biological economy of the sea. Ascending, cold water is rich in the nutrients essential to marine life. Descending, warm water tends to be barren. The mixing associated with large-scale convergences and divergences is an important element of oceanic circulation systems, often extending to depths of over a thousand meters.

What are the sources of ocean currents? The energy for the movement of vast masses of seawater ultimately comes from the atmosphere and the earth's spin. The planet's spin and the heating and cooling of air masses generate planetary winds. Oceanic circulations derive partly from the action of wind on the surface of the sea, partly from the earth's spin, and partly from temperature differences in water masses. Surface currents are predominantly wind-driven. Trade winds, which are largely a consequence of the earth's east-to-west spin, help produce the great equatorial currents. The north equatorial current of the Pacific flows over 14,000 kilometers west from Central America before being deflected northward to become the Kuroshio off the east coast of Japan (Figure 2). From Japan, the Kuroshio mixes with the cold Arctic waters of the Oyashio as the North Pacific current, which flows east across the Pacific near 40 degrees north latitude. Off the coast of North America, the flow is deflected south as the California current, which eventually rejoins the north equatorial current, thus completing the clockwise circulation in the North Pacific basin. A similar circulation occurs in the Atlantic Ocean and includes the Gulf Stream and the Atlantic equatorial current.

Eddies are also characteristic elements of oceanic circulations. They occur in both surface and deep waters and can be so extensive that they cover thousands or even tens of thousands of square kilometers. Some eddies are remarkably persistent and last several years. They are too transient to be shown on maps of average oceanic conditions, yet can have profound effects on the tiny marine plants and animals that may be transported far beyond their typical ranges.

Tropical and subtropical zones tend to have high air pressures, weak winds, clear skies, and much sun. Such factors make surface waters warmer and saltier than those in cooler climates. Cool waters off the coasts of Asia and North America, where the minute plant life called phytoplankton is abundant, are usually gray-green. In contrast, tropical and subtropical seas contain few algae and have a blue cast caused by the scattering of light by molecules of clear water. Biologists have difficulties in defining precisely the difference between tropical and subtropical oceanic zones because they are fairly similar. Tropical zones are usually defined to include all waters where sea surface temperatures remain 23 degrees centigrade or above all year. Tropical zones have only minor seasonal temperature changes in surface waters. Warmer surface waters are separated from cooler subsurface waters by sharp discontinuities called thermoclines, which are present year round and limit the flow of nutrients into surface waters from the deep. The lack of nutrients results in low productivity, especially in areas far from land. Seas adjacent to continents often receive land-based nutrients from rivers, which increase the growth and development of marine organisms.

Some Distinguishing Features of Hawaiian Waters

The waters that stretch to the horizon from the eight main islands (Hawaii, Maui, Kahoolawe, Lanai, Molokai, Oahu, Kauai, and Niihau) are somewhat different from those of the Northwestern Hawaiian Islands, which extend from Nihoa to Kure Atoll. Like oceans everywhere, Hawaiian waters are a quilt of microenvironments, and adjacent water masses possess slightly different physical characteristics. Water temperature, salinity, and primary productivity can vary within short distances as a result of vicissitudes in currents or eddies, rainfall, cloud cover, winds, and marine organisms. Because weather changes constantly, the ocean never becomes a truly homogeneous water mass.

Surface water temperatures near the main Hawaiian islands average about 25 to 26 degrees centigrade during summer and two or three degrees lower in winter. The temperatures of the waters of the Northwestern Hawaiian Islands change correspondingly with the seasons, but are always a degree or two cooler than those of the main islands. All Hawaiian waters have a permanent thermocline, which acts as a physical barrier to mixing between the warm surface water lens and the great masses of deep, cold North Pacific water. The thickness of the warm surface layer varies with the location and the season but is usually about 100 meters near the main islands and 70 in the vicinity of the Northwestern Hawaiian Islands.

Surface salinity in Hawaii is typical of that of subtropical waters, ranging from 35.0 to 35.4 parts per thousand. Salinity is slightly higher between Gardner Pinnacles and Pearl and Hermes Reef. On a global scale, the salinity of virtually all oceanic waters falls between 33 and 37 parts per thousand. As one might expect, warm tropical areas with limited rainfall have the highest concentrations of salts, while cold areas with lots of precipitation have the lowest.

Sunlight is an important factor in every marine ecosystem because photosynthesis is the ultimate source of most life. Light penetrates to a depth that is proportional to the turbidity of the water. Because the blue subtropical waters of Hawaii contain fewer algae than California waters, light penetrates fairly deeply. The photic zone, or layer of water that receives sufficient sunlight to enable photosynthesis, averages more than 100 meters. It is slightly shallower in the Northwestern Hawaiian Islands and decreases during summer, when more phytoplankton are in the water column. In productive areas of the world the photic zone rarely extends more than thirty meters and in extreme situations can be as shallow as a few centimeters. The length of the day varies less throughout the year in Hawaii than in high latitudes. The longest day in Honolulu lasts between thirteen and fourteen hours and the shortest about eleven, while comparable day lengths in Maine are almost sixteen and less than nine. One consequence of fairly uniform day length is that photosynthesis can occur all year, whereas it cannot in such northern waters as the Bering Sea. The growth of algae in Hawaii is never limited by an absence of sunlight.

Permanent thermoclines typically occur in subtropical open-ocean areas where surface layers are strongly heated during most of the year. Surface waters in higher latitudes receive less solar heat and develop shallow thermoclines only during summer, if at all. Winter winds and waves in northern waters mix surface and deep waters sufficiently to break down any temporary stratification that may have developed between them. The permanent thermocline in Hawaii is a barrier to the upward movement of cold, deep subsurface waters that have high concentrations of essential nutrients. Such an impediment to the free flow of nutrients has important consequences for marine organisms.

As in all tropical and subtropical marine habitats, the productivity of Hawaiian waters is limited by the availability of nitrates. Surface waters have similar concentrations throughout the archipelago, but slight increases are apparent during winter and in the Northwestern Hawaiian Islands. Nitrates and to a lesser extent phosphates are scarce in surface waters because they are constantly removed when marine creatures die. Dead plants and animals sink below the thermocline to deep waters, where the nutrients they assimilated in life are released as they decompose. Most surface waters in Hawaii are deficient in nutrients because the lens of warm surface water rarely mixes with the cooler, nutrient-rich deep water. Inshore waters such as the shallow lagoons at French Frigate Shoals and Pearl and Hermes Reef have higher concentrations because nitrates tend to be retained and recycled in the absence of a thermocline. Seabirds may concentrate nitrates in nearshore waters and coral reefs. They feed over vast areas of the ocean, eating prey whose bodies are full of nitrates. When the birds defecate on the islands, their nutrient-rich guano is eventually washed by rain into nearshore waters.

Some areas are enriched by upwellings, eddies, and fronts. On a global scale, locations where upwellings bring nutrient-rich deep water to the surface can be ten times as productive as otherwise similar waters. Enriched waters are often sites for important fisheries such as those found in the Humboldt current off the west coast of South America and the Benguela current off the coast of Namibia, where currents and winds drive surface waters away from the coasts and cold, deep water rises to replace them. On a localized scale, eddies and upwellings are common on the leeward sides of islands and over submerged banks, where deep-water currents collide with submerged volcanic ridges. Uncharted fronts occur in tropical and subtropical waters where water temperature and depth change abruptly. Fronts are associated with increased water turbulence and upwelling, which produce localized concentrations of plants and animals so abundant that the water can have the consistency of soup. In *The Arcturus Adventure*, William Beebe described such a phenomenon in the equatorial Pacific midway between Panama and the Galápagos:

> Again and again I was impressed with one outstanding feature of the Current Rip, this uncharted zoologists' paradise—the narrowness of its limits and the sharpness with which these limits were defined. It was a world, not of two, but to all intents and purposes, of a single plane—length. From first to last we followed its course along a hundred miles, and yet ten yards on either side of the central line of the foam, the water was almost barren of life. The thread-like artery of the currents' juncture seethed with organisms—literally billions of living creatures, clinging to its erratic angles as though magnetized.

Fronts, upwellings, and convergences may greatly enhance the productivity of Hawaiian waters. They are difficult to detect by routine scientific sampling because they are small and seem to occur randomly. During the 1980s biological oceanographers began to believe that the subtropical waters north of Hawaii might be two or three times as productive as earlier researchers had estimated because their sampling ignored such isolated pulses of high productivity. Production varies in the water column and increases substantially near the thermocline, just above the cold, nutrient-rich waters.

The Food Chain

Floating microscopic phytoplankton forms the base of food production in the ocean. These single-celled algae use carbon dioxide, water, and energy from sunlight to photosynthesize carbohydrates. Analogous to green plants on land, they are the primary producers in the marine food web. Because of nitrate limitations, primary productivity in Hawaii is a low 330 metric tons per square kilometer per year, although recent studies imply that this may be an underestimate. Productivity is somewhat higher in summer than in winter and is significantly higher in the Northwestern Hawaiian Islands than in the main islands. It is also greater in waters close to the islands than in offshore waters: the lagoon at French Frigate Shoals produces almost ten times as much phytoplankton as an equivalent volume of water in the open ocean.

The annual primary productivity of the open ocean near Hawaii may be only one-third of that in temperate coastal zone areas and one-tenth of that in open-ocean upwellings. In contrast to the fairly constant levels of productivity of Hawaiian waters throughout the year, production poleward of the tropics and subtropics varies dramatically by season. In arctic and temperate waters phytoplankton suddenly proliferates in spring, providing a vast supply of food for the marine ecosystem. Hawaii has no such spring bloom. Some estuarine and coral reef waters are so productive that in two days they can outproduce the entire annual yield of an equivalent water mass in Hawaii.

The smallest of the creatures that make up plankton, from protozoans to larval fishes, graze on the algae. These primary consumers are in turn eaten by secondary consumers, either carnivorous zooplankton or fish. The position of any species in a food web—whom it eats and who eats it—defines its trophic level. Plants are the first, herbivores the second, primary carnivores the third, and so forth. Tropical and subtropical waters are usually considered to have five trophic levels, the fifth including the largest predatory fishes, such as sharks and tunas.

The five-trophic-level model for Hawaiian waters generally describes the interactions among its thousands of marine plant and animal species, but the actual situation is far more intricate. Trophic levels are more poorly defined in the ocean than they are on land. A predator may feed on three trophic levels during the same day. At different stages in its life cycle, a fish may occupy several trophic levels, consuming zooplankton as a fry and large predatory fishes as an adult. A blue-gray noddy can consume a larval blue marlin and, if unwary, can itself become a snack for its prey's parent minutes later. Interactions among the 11,000 species of fish, 6,000 species of crustacea, and 700 species of squid in the world's oceans can be vastly complex.

Most blue offshore waters in Hawaii are an impoverished biological desert where food chains tend to be long, complex, and inefficient. Each transfer of energy between trophic levels in Hawaii squanders nine-tenths of the original energy, in part because predatory animals spend much more time and energy searching for food than their counterparts in more productive waters. As a result, of every 10,000 units of energy produced by Hawaiian phytoplankton, only one unit winds up in a yellowfin tuna after four energy transfers through five trophic levels. In contrast, productive areas off the coast of California with higher ecological efficiencies and shorter food chains produce four hundred times more energy to top-level predators, such as large fishes, seabirds, porpoises, and humans. At localized upwellings or fronts, Hawaiian productivity may approach California levels. Obviously such oases of food and energy are especially important to the Hawaiian ecosystem and are sought out by hungry predators.

An important phenomenon in the Hawaiian food chain is the daily vertical migration in the water column of planktonic creatures. Unlike phytoplankton, zooplankton is not restricted to the sunlit upper hundred meters of the ocean, but can be found at all depths. Many of these creatures undertake extensive migrations, usually moving toward the surface during the night and descending hundreds of meters during the day. Such movements can be observed as deep scattering layers on ships' echo sounders, though frequently the sophisticated electronic equipment actually detects a school of lanternfish, hatchetfish, or bristlemouths pursuing zooplankton rather than the zooplankton per se. Crustaceans and squid move similarly in the water column. Migrations enable the organisms to feed in the productive euphotic zone yet hide in deep water during the day, when the risk of becoming a meal is greatest. Vertical migration seems to be an important means by which food is transported from the surface to the deep ocean. Deep-water animals consume prey near the surface which otherwise might have remained there to be recycled in surface waters. When the predators descend, they are hunted by resident deep-water creatures. Vertical migrations have influenced the feeding habits of several Hawaiian seabirds.

Some Marine Fauna That Are Important to Hawaiian Seabirds

Only a few of the 700 species of fish and thousands of other marine organisms that live in Hawaiian waters are directly important to foraging Hawaiian seabirds. The amount of fish larvae in surface waters seems to influence seabird life cycles. Fish larvae, especially those of inshore and reef species, are most abundant in surface waters during summer, an indication that fish in Hawaii tend to spawn during the summer months. Larvae are almost twice as abundant in the Northwestern as in the main Hawaiian islands. Common larval fishes include anchovies, lizardfish, skipjack tunas, yellowfin tunas, flyingfishes, goatfishes, lanternfishes, sauries, and dolphinfishes. Lanternfish larvae are particularly abundant at night, an indication that even the young of this family migrate vertically in the water column. Anchovies and Pacific sauries are found primarily in the northwestern portion of the Hawaiian archipelago. Such differences in fish fauna between the cool, subtropical northwest waters and the warmer, tropical southeast waters emphasize the transitional nature of Hawaiian waters. North of Kure Atoll the waters become too cool to support coral reefs.

The nine flyingfish species in Hawaii are of special interest because they are commonly consumed by many Hawaiian seabirds. Flyingfish have unusually large pectoral and central fins that can be used as wings, allowing them to skim hundreds of feet above the ocean's surface to flee from porpoises, tunas, dolphinfish, or swordfish. This means of escape from the mouths of predators from below increases their vulnerability to hungry seabirds, which capture flyingfish near the surface or in mid-air. Errant flyingfish occasionally propel themselves onto the decks of ships. Flyingfish attach their eggs to flotsam such as seaweed, pumice, fishing line, and plastic. Most species breed during winter near the equator, but some lay eggs at least as far north as Midway. All flyingfish are migratory. They seek warm water, and their distribution is limited by the temperature of the surface water. Flyingfish are found in all tropical seas and their presence is almost diagnostic of subtropical or tropical waters. They are common at 23 and abundant at 25 degrees centigrade, moving in winter toward the equator and in summer toward the poles, where they feed on crustaceans such as copepods and amphipods. Their migrations bring many of them to Hawaiian waters during summer but far fewer are found in winter. Larval flyingfishes are common in Hawaii all summer.

All three species of mackerel scads in Hawaii are eaten by Hawaiian seabirds. These spindle-shaped fishes are resident, remaining in Hawaiian waters throughout the year, often schooling in large numbers. Young fish shoal near the reef and shoreline, while adults move into deeper water offshore. Mackerel scads usually feed on zooplankton such as amphipods, copepods, and juvenile forms of pelagic crabs. They fall prey themselves to migratory skipjack tunas and are an important source of food to the tunas during summer. Probably the pressures of predation from tunas drive schools to surface waters, where seabirds are more likely to encounter them. Spawning occurs from spring to late summer, peaking from May to July, when fish fry are most in demand by fledgling seabirds and tunas.

Ommastrephid squids are a surface-dwelling family that are common and widespread in all warm oceans. These long, slender mollusks are migratory, traveling thousands of kilometers into warming waters during summer, then retreating toward the equator in winter. Some species in Japan migrate toward

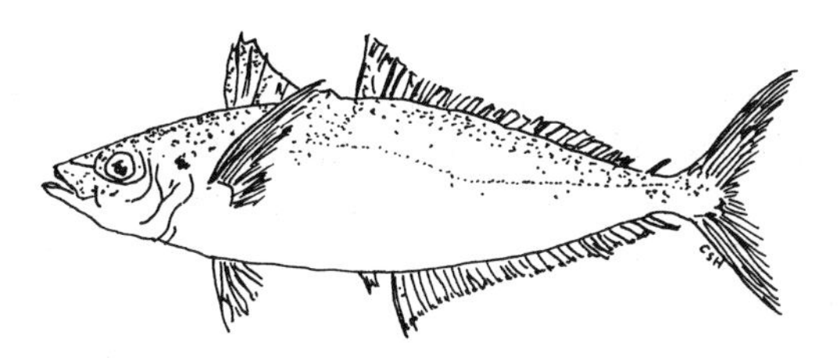

Mackerel scad

shore during April and May to lay eggs on the ocean floor. The migrations and precise breeding seasons of the four common ommastrephid squid in Hawaii are poorly known. Squid produce vast numbers of young, which usually mature within two years. The great frequency with which seabirds, tunas, and porpoises eat juvenile ommastrephid squid implies that squid are a very important component of the Hawaiian marine ecosystem, despite the fact that fishermen and scientists have difficulty catching them. Ommastrephid squid are rapacious and use their hard beaks to attack and devour fish their own size. They also take copepods and small fishes, but they shred their food so thoroughly that it is difficult for biologists to learn the specifics of their diets. Two of the common squid in Hawaii, *Ommastrephes bartrami* and *Sthenoteuthis oualaniensis*, are called flying squid because of their ability to leap from the water as though propelled by submarine cannons when they wish to avoid submerged predators.

Tunas' extensive migration in the Pacific basin poses special problems in international fishery management. Three species are of particular interest to Hawaiian seabirds—skipjack, yellowfin, and little tunas. As we shall see in later chapters, many Hawaiian seabirds feed in association with tuna schools. Tunas forage during the day and apparently make prey available to birds by driving it to the surface: flyingfish and flying squid leap from the water to avoid the tunas. Offshore, skipjack tunas are associated most frequently with bird flocks. The tendency of adult yellowfin tunas to feed well below the surface in Hawaii breaks the linkage with seabirds, but Hawaiian birds occasionally feed with schools of juvenile yellowfin on the surface. Nearshore, bird flocks sometimes congregate over schools of nonmigratory little tunas. Little tuna schools remain fairly close to the shoreline and only rarely venture far offshore. Migratory tunas are most abundant in Hawaii during spring and summer, and fishermen land large amounts of pelagic tunas then. Larval forms are common in Hawaiian waters during summer but rare in winter. Apparently tunas spawn and larvae are present only when the water is 23 degrees centigrade or above. Where waters are cooler, they rarely spawn. The number of tuna larvae near the Northwestern Hawaiian Islands declines from French Frigate Shoals to Midway in concert with declining surface water temperatures.

All marine wildlife depends on the marine ecosystem for sustenance. The natural characteristics of the physical environment provide the ground rules under which all life forms compete to survive. Tropical marine creatures, unlike those in cool northern waters, have the additional challenge of surviving in an impoverished environment that is limited by a lack of nutrients at the base of the food web. Food supplies tend to be patchy, and species that exist millennium after millennium must adapt to take advantage of occasional feasts and survive frequent famines. Seabirds live much of their lives above and in the ocean in search of food. But they are terrestrial expatriates, having originated on land and, like marine turtles and monk seals, must return to terra firma to reproduce. Seabirds have never evolved into true marine creatures because they have never successfully created floating nests at sea. The tie to land is their Achilles' heel.

3 THE HUMANS

High islands have existed in the Hawaiian-Emperor archipelago for as long as 70 million years. Each island arose from the ocean floor, enjoyed a few million years of existence after the lava ceased to flow, then eroded to become an atoll. Various plants and animals arrived in the hit-or-miss fashion characteristic of oceanic islands—an insect blown from Asia by a storm, a seed carried adrift by the ocean from the Gulf of Panama, a spore attached to a shearwater feather. Some species established populations, which over time diverged genetically from their parent stocks. Although no one can be certain, the Northwestern Hawaiian Islands probably served as stepping-stones for some of the flora and fauna that arrived at today's main Hawaiian islands. Many of the creatures that dwelled on Laysan when it was a high island undoubtedly did not reach the younger islands to the southeast but became extinct when Laysan eroded and chance did not provide opportunities for dispersion. Many of today's native species on the main Hawaiian islands appeared within the past few million years, long after Laysan had metamorphosed into an atoll, and their arrival had no direct connection with the older, drowned islands. But the creature that was to transform the islands dramatically was absent during most of the 70 million years. When *Australopithecus* was stalking baboons and antelope on the East African savannah a million years ago, most of the main Hawaiian islands had long since emerged. The high-island stages of the Northwestern Hawaiian Islands had come and gone tens of millions of years before the age of humans.

Before Western Contact

During the declining days of the Roman Empire and just before the Middle Ages in Europe, no human eye had gazed upon any Hawaiian island. No pig or any other cloven-hoofed mammal had trod upon Hawaiian soil. On the eve of colonization by humans, all the land area of the high islands was well forested, except for recent lava flows and the areas above the tree line on the highest mountains. The only land mammal was a small bat. There were no rats, mongooses, dogs, or mosquitos. In an environment where the only predators were hawks and owls, seabird colonies were probably as abundant on the main Hawaiian islands as they are today on the offshore stacks and islets. The arrival of humans on the main islands transformed their ecology, but the effects in the Northwestern Hawaiian Islands were delayed another thousand years.

Polynesian settlements on Oahu date back to at least A.D. 600, and there are reasons to believe that the first humans arrived in their outrigger canoes as early as A.D. 400. They brought with them exotic plants and animals from the South Pacific: coconuts, sweet potatoes, taro, bananas, dogs, pigs, chickens. Polynesian rats probably arrived simultaneously as stowaways on canoes. Like prehistoric peoples on islands elsewhere, the Polynesians had a tremendous impact on the pristine ecosystem of the main Hawaiian islands. As the population of the colonists increased and expanded, the lowland habitats of the main Hawaiian islands suffered extensive alteration and destruction. Endemic bird populations became extinct when their unique habitats disappeared.

Much of the alteration to the lowland ecosystems came from large-scale agricultural operations, which provided much of the food for ancient Hawaiians. The slopes of Halawa Valley, Molokai, were denuded up to elevations of 300 meters by prehistoric slash-and-burn land clearing. Southwest Molokai, Kahoolawe, Lanai, and Niihau were completely barren of trees when Westerners first described them in the late eighteenth century, an indication of extensive land-clearing activities. Early Western reports from Waimea Bay, Kauai, indicate that the plains in that area were periodically burned and forests were so diminished that wood was scarce. Plantations extending ten kilometers or more inland from Kealakekua Bay in western Hawaii may have typified the extent of agriculture on all main islands. Besides extensively altering natural habitats along the coasts of the main islands, the early Polynesians introduced animals that decimated native birdlife. Rats and dogs attack and eat any bird that nests on or burrows in the ground. Pigs, which were abundant by the time Westerners arrived, root out and extirpate entire shearwater colonies.

The Polynesians also contributed to the eradication of many species of seabirds by eating seabird adults, young, and eggs. Bonin petrels and a recently discovered but still unnamed petrel once occupied the main Hawaiian islands but are now extinct there. Before the arrival of humans, wedge-tailed shearwaters undoubtedly nested in the coastal areas of all of the main Hawaiian islands. George C. Munro, a twentieth-century Hawaii ecologist, wrote that Newell's shearwaters were eaten by Hawaiians at Waipio Valley on the island of Hawaii. Bulwer's petrel chicks were considered to be a great delicacy by ancient Hawaiians. Harcourt's storm-petrel remains are common in middens on the island of Hawaii, where the species is now almost extinct.

Although dark-rumped petrels have been exterminated on Oahu, the excavation of middens there indicates that dark-rumps were once the islands' most abundant seabird. Hawaiians so favored dark-rumped young that they were reserved for *alii* (the nobility). Adult birds were also eaten, but they were salted down to mask their strong fishy flavor. Large numbers of dark-rumped petrel bones have been found in the saddle portion of the Big Island between Mauna Loa and Mauna Kea, where old Hawaiians at the turn of the twentieth century remembered vast breeding grounds in the lava. Hawaiians netted many birds along the mountain ridges as dark-rumps returned to their colonies at nightfall. Sometimes smoky fires were lit in the paths of incoming flocks. Like shearwaters or petrels attracted to street lights today on Kauai, the dark-rumps were stunned and disoriented and could easily be captured by hand.

Hawaiians also consumed seabirds on the offshore islets. Manana succumbed so thoroughly to predation by humans that it was devoid of nesting seabirds at the turn of the twentieth century; by the 1970s its protected status had allowed the establishment of large colonies of several species. Ancient Hawaiians occasionally canoed to Kaula to collect seabirds for featherwork.

Storrs L. Olson's studies of fossil birds indicate that at least 39 species that were present before the arrival of Polynesians—more than half of Hawaii's endemic birds—were extinct before Europeans encountered Hawaii. Habitat alteration or direct human consumption probably caused many other species to become restricted in distribution or to disappear from certain islands during the Polynesian settlement. It is no surprise that many of the species that were already severely stressed became extinct soon after Western technology and land use practices added additional environmental perturbations to the surviving remnants of the original habitats. Humans are natural predators, and prehistoric humans in Hawaii were no different from people elsewhere in wreaking destruction on a fragile ecosystem. As the thriving muttonbird industry in Australia and New Zealand attests, young seabirds make excellent eating. Seabird bones have been found in Pleistocene middens in the Aleutians, California, and Europe, and on Ascension and Norfolk islands. Primitive Polynesians have had similar effects elsewhere. For example, Polynesians inhabited Henderson Island in the South Pacific for only about a century, yet managed to drive at least one-third of the birds that once lived there to local extinction, including red-footed boobies, Christmas shearwaters, and white-throated storm-petrels. The Marioris survived on seabirds and fish in the Chatham Islands and almost brought the endemic magenta petrel to extinction. They eventually became prey themselves when the more warlike Maoris invaded and proved to have a taste for human flesh.

The first humans to arrive in the Northwestern Hawaiian Islands were apparently Polynesians of Tahitian origin who, in the thirteenth century, were either forced out of the main Hawaiian islands or, like many nineteenth-century Europeans, shipwrecked on Nihoa or Necker. Although both islands were occupied about the same time, the oral culture of precontact Hawaiians recognized the existence only of Nihoa. Anthropologists who have studied the remains of structures left on Nihoa and Necker have concluded that they were inhabited for relatively brief periods of time following the first Polynesian settlement of the main Hawaiian islands. Stone houses and ceremonial structures are still present on Nihoa, and K. P. Emory of the Bernice P. Bishop Museum in Honolulu estimated that about a hundred people may have lived on the island at one time. Only about twelve acres of Nihoa have gentle slopes, and they are entirely stepped with cultivation terraces. Undoubtedly the inhabitants supplemented their crops with birds, eggs, and anything that they could obtain from the sea. Today there is little fresh water on either Nihoa or Necker, but the situation may have been different during the period of human occupancy.

We can surmise little from the grindstones, bowls, jars, bird bone awls, cowrie squid lures, and human skeletons that have been excavated on Nihoa except that they appear to form a pure sample of Polynesian culture of this era. Life must have been extremely difficult in such terrain. The steep rocky slopes and the difficulty of beaching an outrigger canoe would have made the small volcanic cones particularly claustrophobic. Necker has few hollows that could provide shelter from the summer sun or from winter storms. The large number of stone marae on Necker indicate that a major pastime of the inhabitants must have been prayer for deliverance from their miserable existence to more hospitable living conditions.

What were the main Hawaiian islands like just before Western contact? The coastal areas were well settled, with human populations of between 200,000 and 250,000. Six of the eight main islands had more residents in 1778 than they have now. The mountain forests were probably relatively pristine, but the coastal and lowland areas were highly disturbed by human activities. Some of the changes wrought by humans were beneficial to wildlife: taro ponds created habitat for coots, gallinules, ducks, stilts, and other waterbirds. But many birds were already extinct and others had sharply reduced ranges. Europeans brought many changes to the Hawaiian Island ecosystem, but no more than those wrought by Hawaiians during the first millennium of human occupancy.

Western Discovery and Early Exploration

Captain James Cook, possibly the greatest explorer of the eighteenth century, is believed to have been the first European to set foot on the main Hawaiian islands. During the course of his third major voyage in the Pacific, the Englishman sighted Oahu in January 1778. He visited Kauai and Niihau a few days later, when half a world away George Washington was braving a winter at Valley Forge. A year later, after sailing to North America, Cook returned to Hawaii and was killed by Hawaiians at Kealakekua, on the island of Hawaii. Although Cook's death dissuaded other ships from putting in at Hawaii for several years, geography held the upper hand. By the turn of the nineteenth century, Westerners recognized Hawaii's key position between Asia and North America. The prime location combined with the discovery of sandalwood brought more Westerners, new technologies, and new creatures to Hawaii.

European ships carried a wide variety of animals on their voyages of exploration and settlement. Within a short period, European pigs, goats, cattle, sheep, horses, cats, rabbits, black rats, and Norwegian rats were introduced to the Hawaiian Islands. These alien creatures created additional pressures on the indigenous Hawaiian plants and animals through direct predation and changes in habitat. Because the native flora was poorly adapted to large numbers of grazing and browsing animals, it yielded to exotic plant species. By the mid–nineteenth century, vast forests had been destroyed by grazing animals. The accidental introduction of mosquitos in 1826 brought avian malaria, which killed large numbers of birds at the lower elevations. During the second half of the century, lowland areas were cleared for sugar cane, pineapple, and other crops. Because rats were causing severe damage in cane fields, planters introduced mongooses in 1883 to control them. Unfortunately, mongooses became a more serious threat to ground-nesting birds than to rats on Oahu, Molokai, Hawaii, and Maui. By the turn of the twentieth century, mongooses were occupying burrows of dark-rumped petrels on thousand-meter cliffs on Molokai. Native birds were also exposed to a wide variety of diseases and parasites by the 150 species of exotic birds that were released in the main Hawaiian islands. Many had little resistance to alien diseases and succumbed.

By the turn of the twentieth century, most vegetation below 500 meters in elevation was exotic. Naturalists saw few native birds along the coast or in the lowlands, as about half of the endemic bird species had become extinct under the combined pressures of Polynesians and Westerners. The few seabird colonies that endured in the main Hawaiian islands were restricted to the most isolated areas, where they escaped the reach of rats, mongooses, and humans. Even offshore islets such as Manana Island were devoid of seabirds. These changes occurred despite such a precipitous decline in the human population that Hawaii had only 56,000 people in 1876, less than one-quarter of the number when Cook arrived.

The Northwestern Hawaiian Islands were discovered during the half century following Captain Cook's arrival by navigators flying a variety of flags. Necker and French Frigate Shoals were discovered by the French captain Jean-François de Galaup, comte de La Pérouse, in 1786 as he voyaged from Monterey, California, to Macao. Heavy surf precluded a landing on Necker, which Pérouse named after Jacques Necker, minister of finance under Louis XVI. Two nights later, only a last-minute change of course prevented the *Broussole* from becoming the first of many vessels to run aground on the reef at French Frigate Shoals. The next morning, the captain carefully surveyed the Basse des Fregates Françaises, but La Pérouse Pinnacle was not named after its discoverer until many years later. The Westerners who discovered Nihoa, like those who first saw Necker, did not go ashore. Nihoa was sighted by the British captain William Douglas in 1789. The pounding surf persuaded him not to attempt to land, and his log observes that the island "does not seem to be accessible but to the feathered race."

By the late eighteenth century and early nineteenth, traders began to call in the Hawaiian Islands with increasing frequency. Whalers moved into the northern oceans in the 1820s, and by 1825 Hawaii provided the most important ports in the Pacific. The increased traffic led to further discoveries of the Northwestern Hawaiian Islands. Historical records are sketchy, but a Spanish vessel probably discovered Kure Atoll in 1799. It was named, however, after a Russian captain who visited the atoll on the *Moller* in 1827. Captain Kure may also have been the first to land at Midway Islands, which he visited in 1825, but the records of his journey are incomplete. Lisianski was discovered in 1805 when the *Neva*, under Captain Urey Lisianski, ran aground on a voyage from Sitka, Alaska, to Macao during the first Russian attempt at a circumglobal expedition. The forgiving crew unanimously voted to name the island after their captain. Joseph Allen on the Nantucket whaler *Maro*, a codiscoverer of the famous Japanese whaling grounds, found Gardner Pinnacles and Maro Reef in 1820. No one landed on Gardner Pinnacles before Captain Kure's arrival some eight years later. Pearl and Hermes Reef was discovered in 1822 when two British whalers, the *Pearl* and the *Hermes*, were wrecked on the reef on the same night a mere fifteen kilometers apart. The crews established a camp on one of the islands, managed to salvage enough material to construct the *Deliverance*, and sailed back to Honolulu. Laysan is believed to have been discovered by American whalers in the early 1820s, but details are murky.

Sporadic claims to ownership of the various Northwestern Hawaiian islands were made during the nineteenth century. Queen Kaahumanu sent several vessels to Nihoa to annex it to the Hawaiian kingdom in 1822. Captain John Paty and King Kamehemeha IV voyaged to the Northwestern Hawaiian Islands in 1857, and again claimed Nihoa and annexed Laysan, Lisianski, and Pearl and Hermes Reef to the Hawaiian kingdom. In 1858–59 the U.S. schooner *Fenimore Cooper* carefully charted the islands from Nihoa to Laysan to explore the possibility of laying an underwater telegraph cable from Hawaii to Japan. French Frigate Shoals was formally claimed on behalf of the United States under the Guano Act of 1856. Captain Brooks of the Hawaiian vessel *Gambia* claimed possession of Midway for the United States in 1859. Kure Atoll was ignored until 1886, when King Kalakaua sent an emissary to take possession of it for the Hawaiian government. Title to Necker became an issue in the 1890s, when Great Britain needed a mid-ocean cable station in the Hawaiian archipelago to improve communications between Canada and its Australasian colonies. In order to confirm its status, Captain James A. King, Hawaii's minister of the interior, voyaged to Necker to claim it for the provisional Hawaiian government in May 1894. All of the Northwestern Hawaiian islands were acquired by the United States government in 1900 with the organic legislation that took effect when the Hawaiian Islands became a territory of the United States.

Throughout the nineteenth century numerous vessels were wrecked in the Northwestern Hawaiian Islands. Stranded sailors understandably made use of whatever sustenance the islands and atolls offered. Monk seals and turtles basking on the beaches were unaccustomed to terrestrial predators and made easy prey and possibly good eating. Bird eggs and flesh were available virtually throughout the year and provided additional fare. Some castaways severely affected seabird populations. George C. Munro found no albatrosses, boobies, tropicbirds, or wedge-tailed shearwaters on Sand Island, Midway, in 1891. Apparently shipwrecked sailors during the previous century had ravaged the bird populations. Stranded sailors were lulled into a false sense of abundance. Unaware that very few islands exist in the north-central Pacific, early visitors did not understand that wildlife from hundreds of thousands of square kilometers of ocean concentrate on the only available land during breeding season. Sailors' stories of unlimited wildlife lured others whose interests went far beyond mere survival.

Numerous whaling, commercial, and exploratory vessels plied the waters of the Northwestern Hawaiian Islands during the nineteenth century and scoured the islands and surrounding reefs for anything that could be eaten or sold. Records of many visits are either nonexistent, scanty, or buried in musty logbooks. Fresh seal meat, sealskins, bird down, bêche-de-mer, shark fins, fish oil, and turtle shells were the most frequent commodities removed from French Frigate Shoals, Laysan, and Lisianski. One voyage stands out as particularly rapacious. The Japanese schooner *Ada* was chartered by an American, George Mansfield, during the off season of sea otter hunting. In 1891–92 it sailed from Yokohama to the Bonin (Ogasawara) and Northwestern Hawaiian islands. The crew of the *Ada* took 120 turtles from Lisianski and an additional 130 from Laysan, far more than the entire population of either island in the 1980s. The ship's log records finding a sign on Laysan which "was an appeal to voyagers not to take the turtle away." After taking every turtle they could grab, the crew repainted the board to express identical sentiments. Bird down was another valuable resource the crew collected as they plundered their way down the chain. They obtained albatross down by killing chicks, dipping them in boiling water, and then stripping off their feathers. Pearl and Hermes Reef provided twenty kilograms of down, which probably accounted for every chick on the atoll. It was inevitable that such unregulated commercial pressures would devastate the finite resources of the Northwestern Hawaiian Islands.

During the 1850s, companies throughout the world became interested in locating sites for guano mining in the central Pacific, and the Northwestern Hawaiian Islands did not escape their scrutiny. Significant deposits of guano are typically produced by colonies of albatrosses, boobies, and terns, sooty terns being especially productive. Because boobies and related species such as pelicans and cormorants are prone to deposit most of their droppings at the colony, islands with such species usually produce the most guano. Early reports from the Northwestern Hawiian Islands indicated that several had commercial quantities of guano. From 1855 to 1861 Hawaii supplied a large amount of guano, although the records do not clearly indicate the precise locations of mining activity. There is reason to believe that much of the guano was collected at French Frigate Shoals during 1859. In the 1890s the Hawaiian government granted guano leases for French Frigate Shoals, Laysan, Lisianski, Pearl and Hermes Reef, Midway, and Kure. Among these sites, only Laysan proved to have commercially significant quantities. Guano was dug north of the lagoon on Laysan from 1892 to 1904, during which time as many as forty resident Japanese laborers worked in the mining operation. Clipper ships arrived at the colony full of provisions and departed laden with guano.

Max Schlemmer, the manager of the mine, set loose guinea pigs and two types of rabbits in 1903 with the hope that they would live off the native vegetation and provide a source of fresh food independent of supply ships. It proved to be a classical ecological catastrophe. Within six years the rabbits had overrun the island, consumed the vegetation, created desert conditions, and caused the extinction of three endemic land birds and two endemic plants. Rabbits created similar havoc on Lisianski when they were introduced there. A visitor to Lisianski in 1915 found the only surviving plants to be an introduced tobacco patch and beach morning-glory. The rabbits on Lisianski died out without human intervention, but those on Lasyan were not eliminated until expeditions in 1912–13 and 1923 brought sufficient men and ammunition to finish them off.

Beginning in 1902, Japanese feather hunters visited the Northwestern Hawaiian Islands. The seabird feathers that they collected were used in the millinery trade, primarily for women's hats in Europe and North America. Apparently the Japanese moved into the Hawaiian archipelago only after the seabird populations at Wake, Marcus (Minami Torishima), the Bonins, and the Izu islands had been devastated. Because American law-enforcement agencies rarely patrolled such remote islands, much of the early activities went undetected and we will never know the full extent of the feather hunting. Several times patrol ships found signs of Japanese activities at Pearl and Hermes Reef, but no feather hunters were ever apprehended at that atoll.

Most of the feather poaching apparently occurred on Laysan, Lisianksi, and Midway Islands, probably because they had the largest populations of seabirds and were the farthest from the scrutiny of the main islands. In 1902 a visitor to Eastern and Sand islands, Midway, found the islands thickly strewn with thousands of bodies of wingless and tailess seabirds. During the next two years, parties of Japanese were discovered on Lisianski. One group of 77 was arrested and brought to Honolulu by a revenue cutter, whose crew seized about 300,000 dead sooty terns. At least seven Japanese vessels visited the Northwestern Hawaiian Islands under the pretext of deep-sea fishing during the fall and winter of 1908–9. A revenue cutter arrived at Laysan in 1910 and seized another 300,000 bird wings together with an additional two tons of feathers. Visitors to Laysan during this period reported the carnage to be awesome: portions of the island were devoid of bird life, and bird bones bleached in the sun. The poachers

Albatross eggs, Laysan Island, 1903

used two particularly gruesome techniques. Birds that were not left to bleed to death after their wings had been cut off were slowly starved, a technique that eliminated the fat and facilitated the removal of feathers. Marauding plumage hunters destroyed millions of albatrosses, terns, and shearwaters. The outcry from conservation groups prompted President Theodore Roosevelt to promulgate an executive order in 1909 which created the Hawaiian Islands Bird Reservation. All of the Northwestern Hawaiian islands except Midway were set aside as a wildlife preserve for native birds. Feather poaching continued for several years after Roosevelt issued his order, but patrols by the U.S. revenue cutter *Thetis* resulted in so many arrests that the poachers eventually stayed away.

The Modern Era

The first modern scientific expedition specifically designed to study the natural history of the Northwestern Hawaiian Islands was the Rothschild expedition in the spring and summer of 1891. Walter Rothschild of Tring, England, sent Henry Palmer and George C. Munro to the Northwestern Hawaiian Islands to collect birds. The collections formed the basis of Rothschild's *Avifauna of Laysan and the Neighboring Islands*, which was issued serially in London from 1893 to 1900. The German naturalist H. H. Schauinsland spent three months on

Laysan in 1896 and produced *Drei Monate auf einer Koralleninsel,* which included notes on the birdlife. David Starr Jordan and Barton Evermann supervised an expedition by the United States Fish Commission's *Albatross* in 1902. Those surveys culminated in the three-volume *Aquatic Resources of the Hawaiian Islands,* which included seabird studies by Walter K. Fisher. The *Tanager* expeditions of 1923 and 1924, sponsored jointly by the U.S. Biological Survey, the Navy Department, and the Bernice P. Bishop Museum, added greatly to the store of knowledge of natural history and in addition eliminated the residual rabbit population on Laysan.

Patrols and surveys continued during the 1930s and 1940s, but tended to be geared toward the use of the islands for military and strategic purposes. Many fish and wildlife studies have taken place since the 1950s, to promote both fishery development and wildlife conservation. Since World War II, at least one biological survey has usually been carried out each year. The most concentrated work has been the Pacific Ocean Biological Survey Program during the 1960s, sponsored by the Smithsonian Institution, and the Tripartite Research Program in the 1970s and 1980s, a joint venture by the U.S. Fish and Wildlife Service, the National Marine Fisheries Service, the State of Hawaii Department of Land and Natural Resources, and the University of Hawaii Sea Grant Program. The purpose of the Tripartite Program was to study fish and wildlife biology in order to plan for fishery development and wildlife conservation in the Northwestern Hawaiian Islands.

World War II brought increased activity to the Hawaiian Islands. In the main islands, Manana, Kaula, and Kahoolawe became targets for bombing and strafing practice. The latter two continue to be used as targets today. In the Northwestern Hawaiian Islands, French Frigate Shoals became an exercise location for ships, submarines, and seaplanes during the mid-1930s. After the Battle of Midway in 1942, French Frigate Shoals became an important refueling stop for fighter planes between the main islands and Midway. In the summer of 1942, the navy created a fixed aircraft carrier by dredging a new island at the old location of Little Tern Island. At the same time, an area was cleared of coral heads to allow seaplanes to land. Reconnaissance flights operated out of the naval air station at French Frigate Shoals for the duration of the war, but no battles were fought there. In 1943 the U.S. Coast Guard established a long-range navigation transmitting (LORAN) station on East Island, French Frigate Shoals, which functioned as a navigation aid for ships. It remained on East until 1952, when it was relocated to nearby Tern.

Midway experienced far greater changes than French Frigate Shoals. Beginning in 1903, the Commercial Pacific Cable Company took possession of Sand Island, which was said to be a barren waste of ground coral. A relay station for the submarine cable between Asia and North America brought the first permanent colony. A second commercial enterprise began in 1935, when Pan American Airways established an airport on Sand Island as part of its flying boat service between Manila and California via Guam, Wake, Midway, and Honolulu. A modern hotel, shops, warehouses, and power plants were built, providing most of the amenities of modern life on an atoll in the middle of the Pacific Ocean. The first detachment of marines arrived on Midway in September 1940 and began building defense installations. Midway was set aside as a naval defense area in February 1941.

The naval battle of Midway, a turning point in the Pacific war, served to convert the island into a major military garrison. Collisions with vehicles and aircraft, construction activities, and egg collecting reduced bird populations considerably. The mere presence of 15,000 human beings in such a small area was an important factor in the destruction of many birds. More significant in the long term was the introduction of black rats and mosquitos from ships during 1943. Both Sand and Eastern are now overrun by rats, which have reduced or eliminated several ground-nesting birds. Midway, especially Eastern Island, had substantial numbers of Laysan rails (which had been introduced there by the Rothschild expedition in July 1891) long after they had been eliminated on Laysan. The rails too became extinct within two years of the arrival of rats and mosquitos. Avian malaria, which arrived with the mosquitos, may have contributed to the decline of the rails. After the war, Midway remained an important base for refueling and servicing ships and aircraft. Although the numbers of humans residing there dropped to about 3,000 soon after the war, thousands of albatrosses and other birds were killed by collisions with aircraft, antennas, and guy wires.

Several commercial fisheries have operated in the Northwestern Hawaiian Islands during the modern era. Pearl oysters were discovered at Pearl and Hermes Reef in 1928, and several tons of pearl shells were sold to button manufacturers. The fishery lasted only a few years. Various fisheries operated with varying degrees of success in French Frigate Shoals during the late 1940s and the 1950s; most of them were short-lived. During the 1970s and 1980s, fisheries expanded in the Northwestern Hawaiian Islands, including viable fisheries for spiny lobster, deep-sea shrimp, and tunas. No fisheries operate within any of the lagoons.

Most of the Northwestern Hawaiian Islands are now managed as a federal wildlife refuge by the U.S. Fish and Wildlife Service. Nihoa, Necker, French Frigate Shoals, Maro Reef, Laysan, Lisianski, and Pearl and Hermes Reef have been declared research natural areas within the National Wildlife Refuge System, and visits to these fragile islands and atolls are authorized only for scientific or official purposes. President Harry S. Truman signed an executive order in 1952 which placed Kure Atoll under the jurisdiction of the Territory of Hawaii, and it is now managed as a wildlife refuge by the State of Hawaii Department of Land and Natural Resources as part of its seabird sanctuary. Curiously, the only Northwestern Hawaiian island that is controlled by the State of Hawaii is the farthest from the main islands.

Since 1961, Kure Atoll has supported a U.S. Coast Guard LORAN station with about twenty personnel. The LORAN station at Tern Island, French Frigate

Shoals, was abandoned in 1978 and has since been occupied for research and management by the U.S. Fish and Wildlife Service. Midway is a territory of the United States and as such is not legally part of the State of Hawaii or the United States of America. Midway continues to function as a U.S. Navy air facility yet is managed as an overlay refuge by the U.S. Fish and Wildlife Service. Only a few hundred people live on Sand and much of the base has an air of abandonment. All personnel were withdrawn from Eastern soon after the war, and today the buildings and runways there are dilapidated. Most the islets offshore the main Hawaiian islands are managed as wildlife sanctuaries by the Department of Land and Natural Resources. Administratively, all of the Northwestern Hawaiian islands and islands offshore the main Hawaiian islands have been included within the City and County of Honolulu since 1907. On its face, it seems strange for an urban county to stretch 2,000 kilometers across the Pacific, encompassing volcanic islets renowned for their wildlife. However, the situation is not unique. Metropolitan Tokyo includes the Ogasawara (Bonin) Islands, some thirty islets known for their marine wildlife, which stretch 1,500 kilometers south from Japan.

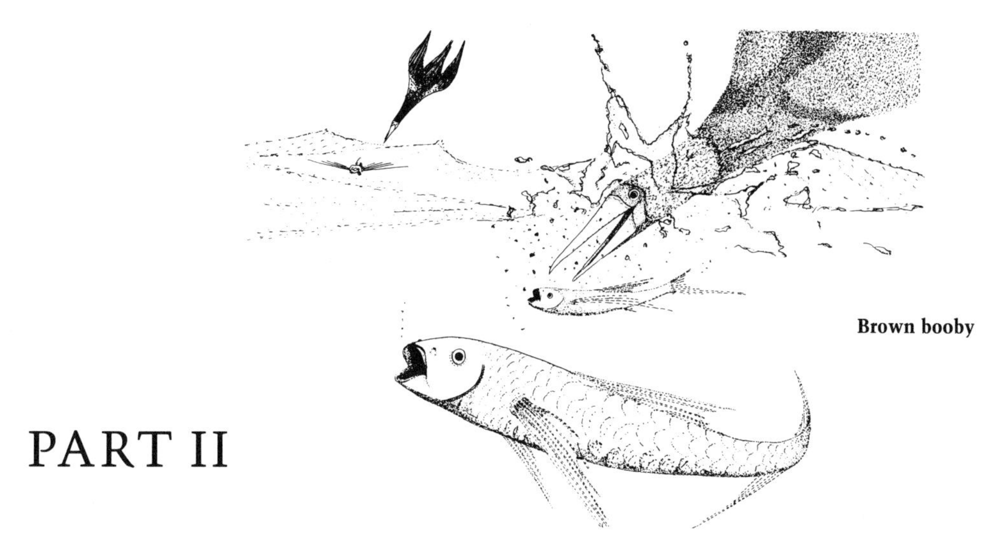

Brown booby

PART II

Comparative Biology of Hawaiian Seabirds

This morning I saw a frigatebird, which makes boobies vomit what they have eaten and then catches it in mid-air. The frigatebird lives on nothing else, and even though it is a seabird, it does not alight on the water and never is found more than 20 leagues from land. There are many of them in the Cape Verde Islands. A little later I saw two more boobies and many flyingfish. They are about a foot long and have two little wings like a bat. These fish sometimes fly above the water at about the height of a lance, rising in the air like a harquebus shot. Sometimes they fall on our ships. The breeze is sweet and delightful. The sea is as smooth as a river, only the nightingales are lacking.

—LOG OF CHRISTOPHER COLUMBUS, September 29, 1492

4 ORIGIN AND ADAPTATIONS OF HAWAIIAN SEABIRDS

There is something ancient and mysterious about great frigatebirds soaring on a thermal above Oahu's Kawainui Marsh or a flock of masked boobies gliding home in V formation during the late afternoon over a Laysan beach. The watchful eyes of similar creatures once surveyed coastal panoramas of woolly mammoths, saber-toothed tigers, giant sloths, dawn horses. Biologists who work with frigates and boobies sometimes refer to them as pterodactyls, acknowledging that they are as much living relics of the past as elephants, rhinoceroses, and monitor lizards.

The sea is the true home of Hawaiian seabirds and to survive they have adapted to a saltwater world. Unlike the mythical halcyon, however, no seabird can build a floating nest on the ocean and raise young there. The inability to breed on their feeding grounds has required Hawaiian seabirds to adapt to a second environment: tropical and subtropical islands. The sea is the oldest and simplest habitat on earth. Covering five-sevenths of the earth's surface, it is also the largest, and one might think that it would be the normal habitat and principal source of food for a substantial number of the world's 9,000 species of birds. But only about 260 species of seabirds exist today, if we eliminate loons, grebes, and sea ducks, which are sometimes found in saltwater but which are truly denizens of continental lakes, rivers, and lagoons. It is obvious that seabirds are not nearly so successful as land birds in terms of speciation. The relationships among the Hawaiian seabirds that exist today and the means by which they have adapted to life in a warm marine environment are now beginning to be understood.

The Evolution of Seabirds

Most biologists agree that life began at sea. About a billion years ago the first unicellular animal life appeared in marine waters, followed some 500 million years later by the first vertebrates. Powered flight, the hallmark of birds, actually evolved independently in three different vertebrate groups. Birds may have been the first to develop the ability to fly some 225 million years ago. Flying reptiles, popularly known as pterodactyls, are known from at least 200 million years ago in the late Triassic and died out about 70 million years ago. The third group to take to the air was mammals, in the form of bats, about 50 million years ago.

Until recently, the earliest fossil bird recognized as such by science was *Archaeopteryx lithographica*. It was found in southern Germany over a century ago and dated back 150 million years. *Archaeopteryx* has so many characteristics of both reptiles and birds that scientists who have studied the five known specimens agree that it is fortunate that the feathers were preserved or they might not have recognized it as a bird at all. This land bird apparently lived in trees near lagoons or shallow seas and is closely related to a group of small two-legged dinosaurs. In 1986 *Protoavis*, a crow-sized ancestor of modern birds dating from 225 million years ago, was discovered in Texas. More birdlike than *Archaeopteryx*, it still retains some characteristics of dinosaurs, such as teeth, long tails, pelvises, and hind legs. Birdlike features include hollow bones, wide eye sockets, and deep, keellike breastbones to which powerful, wing-flapping muscles were attached. *Protoavis* strengthens the close evolutionary link between dinosaurs and birds, which is so well recognized that the distinguished vertebrate anatomist Alfred S. Romer referred to birds as feathered reptiles.

Paleontologists have discovered fossil remains of three orders of extinct marine birds. The fossil record is far from complete for much of the early evolution of birds, and important discoveries such as *Protoavis* will probably continue to be made. Two of the orders are intermediate between *Archaeopteryx* and modern birds. *Hesperornis*, or dawn bird, and its relatives lived toward the end of the age of reptiles, 80 to 100 million years ago. Even the oldest seamounts in the Hawaiian-Emperor chain probably had not begun to form undersea volcanos when dawn birds flew the skies. Their jaws had true teeth, a feature that immediately distinguishes them from modern birds. Dawn birds were large, some two meters in length, and probably were shaped like modern loons. With vestigal wings and powerful legs placed toward the back of their bodies, they were perfectly adapted for diving. These flightless birds probably lived by consuming fish in warm marine environments. *Ichthyornis*, or fish birds, and their relatives have been found in the same late-Cretaceous marine sediments as dawn birds. Their bills also possessed true teeth, probably to grasp the slippery fish that were their prey. Unlike dawn birds, fish birds had good powers of flight and possessed skeletons similar to those of modern flying birds. Robin-sized, they had short legs and probably resembled modern terns or gulls. Such similarities are attributed to convergent evolution because fish birds were not predecessors of either terns or gulls. *Odontopteryx,* or bony-tooths, and their relatives lived between 70 and 3 million years ago and have been recovered from tropical sediments. Unique bony, toothlike projections on both upper and lower jaws gave their bills a serrated appearance. Bony-tooths had long wings similar to those of albatrosses, booby-like bills, and short, stout legs. Such birds could have occurred in Hawaiian waters and nested on many of the older islands and seamounts in the Hawaiian-Emperor chain.

The four orders of marine birds in existence today diverged from their terrestrial ancestors long ago. The Sphenisciformes (penguins), Procellariiformes (albatrosses and their relatives), Pelecaniformes (pelicans and their relatives), and Charadriiformes (gulls, terns, and their relatives) were well-differentiated orders 60 million years ago, long after Meiji Guyot, the northernmost seamount in the Hawaiian-Emperor chain, had emerged. A fossil tropicbird some 60 million years of age and a fossil frigatebird some 50 million years of age have been found, and the pelecaniform ancestor to them both must have diverged much earlier. Over time, the birds within each order diverged through the processes of natural selection to fill the various niches that were available in the rather simple, warm marine environment in Hawaii. In order to reproduce successfully, smaller or weaker birds had to breed at different times of the year, use poorer or different nesting sites, or use smaller or different food sources. Changes in wing proportion, bill structure, body size, and other features created the means by which different types of birds could exist simultaneously. Within each order the differences can be extreme. A 2,800-gram black-footed albatross and a 43-gram Harcourt's storm-petrel have a common ancestor. Sooty terns remain on the wing for years at a time, while blue-gray noddies must return to roost each night. The rate of radiation into new families and genera intensified about 35 million years ago, when the albatrosses, petrels, and shearwaters speciated and the booby family became established. Almost all of the orders and families of seabirds originated in the Indian and Pacific oceans and radiated from there, establishing species in virtually every oceanic zone. The Southern Hemisphere accounts for two-thirds of the world's ocean area, and it is not surprising that most marine bird species originated there.

Penguins

All sixteen species of penguins are restricted to the Southern Hemisphere, where they are commonest in the cooler waters of the southern ocean. No penguins live now or have ever lived in Hawaii. The Galápagos and Peruvian penguins live near the equator and can be considered tropical species. However, both species inhabit waters that are cooled by the Humboldt current and are normally temperate. Penguins abandoned flight many millions of years ago and move by porpoising through the water, often at great speed. Their wings are highly adapted for swimming and, unlike those of most birds, lack flight feathers and cannot be folded. Penguins differ from other seabirds in having a layer of blubber, a double layer of feathers to protect against heat loss, and physiological adaptations for deep diving.

Procellariiformes

The order Procellariiformes, or tubenoses, consists of four families—albatrosses, shearwaters and true petrels, storm-petrels, and diving petrels. Hawaii has no diving petrels. Three of the thirteen living species of albatross occur in Hawaii. Black-footed albatrosses and Laysan albatrosses are common breeding species in the Northwestern Hawaiian Islands and short-tailed albatrosses now make regular landfalls there. All three species range over the entire North Pacific during summer. Many breeding colonies of black-footed, Laysan, and short-tailed albatrosses have been eliminated in the North Pacific during the past century. Short-tails are an endangered species and in the late 1980s numbered only about 400 individual birds.

Of the fifty-five species of shearwaters and true petrels in the world today, six breed in Hawaii. Christmas shearwaters are fairly common in Hawaii and breed elsewhere at several locations within the tropical and subtropical Pacific Ocean. Wedge-tailed shearwaters are abundant in Hawaii and in numerous colonies throughout the warm waters of the Indo-Pacific region. Newell's shearwaters, in contrast, are known only from the main islands of Hawaii and are considered a threatened species. Bulwer's petrels are locally abundant in Hawaii and also can be found in many colonies scattered in the warm waters of the Atlantic and Pacific oceans. Bonin petrels are also locally abundant in the Northwestern Hawaiian Islands and breed elsewhere only in the Bonins (Ogasawaras). Dark-rumped petrels nest only in Hawaii and the Galápagos Islands and are endangered.

Two of the twenty storm-petrel species that are recognized by taxonomists breed in Hawaii. Harcourt's storm-petrels, also known as Madeiran storm-petrels, breed in small numbers in the main Hawaiian islands and are considered an endangered species by the State of Hawaii. They are found in the warm waters of the Atlantic and Pacific oceans, but in the Pacific are restricted to Hawaii, the Galápagos, and Hide Island, Japan. Sooty storm-petrels, also called Tristram's storm-petrels, breed in fair numbers at several of the Northwestern Hawaiian islands. Elsewhere, they are found only in the Volcanos (Iwos) and southern Izus of Japan, where they are reported to be declining. Because of the small population and generally uncertain status of sooty storm-petrels, the U.S. Fish and Wildlife Service considers it to be a sensitive species and monitors its status for possible protection under the federal Endangered Species Act.

Pelecaniformes

The order Pelecaniformes consists of six families—tropicbirds, pelicans, boobies, cormorants, anhingas, and frigatebirds. Hawaii has no pelicans, cormorants, or anhingas. Two of the world's three extant species of tropicbirds breed in Hawaii. Red-tailed tropicbirds breed throughout the Hawaiian Islands and are abundant. Their colonies are widespread throughout the warm waters of the Indo-Pacific region. In Hawaii, white-tailed tropicbirds are generally restricted to the main Hawaiian islands. They are widespread and abundant elsewhere and can be found in the tropical and subtropical waters of all the world's oceans. Red-billed tropicbirds are sometimes seen in Hawaiian waters as stragglers, but none nest there. Only one of the world's five species of frigatebirds breeds in Hawaii. Great frigatebirds are common breeders throughout the Northwestern Hawaiian Islands and a few nest in the main islands. They range widely over the tropical Indian and Pacific oceans and nest at a few islands in the South Atlantic. Some lesser frigatebirds have been sighted in the Hawaiian chain, but they are not known to nest there.

Nine species of boobies, which are called gannets in cold-water regions, exist today. Three boobies nest in Hawaii. Masked boobies, also known as blue-faced or white boobies, breed on every island group in the Northwestern Hawaiian Islands and a few pair breed at Lehua and Moku Manu, in the main islands. They are widespread throughout the subtropical and tropical waters of the Atlantic, Indian, and Pacific oceans and breed throughout their range. Brown boobies are the least abundant of the boobies in Hawaii, but small groups live on almost every island group in the Northwestern Hawaiian Islands and a few breed in the main islands. Brown boobies are considered to be the most common boobies in the tropical and subtropical belts of the world's oceans, but this belief may stem from the fact that their inshore habits make them conspicuous beyond their numbers. Despite their wide range, their populations in many areas are small and threatened, especially in the Indian Ocean, where they are harvested for human consumption. Red-footed boobies are the most numerous of the boobies in Hawaii and breed throughout the archipelago. They are common in all warm waters of the world and breed at seventy or more locations throughout the Pacific, Indian, and Atlantic oceans.

Charadriiformes

The order Charadriiformes includes sixteen families, but only the family that includes gulls, terns, and noddies is of primary interest here. This order includes other families: jaegers, auks, stilts, sandpipers, and plovers. Some species of the latter families migrate to Hawaii during winter but none breed there. Most indigenous nonbreeding visitors to Hawaii are Charadriiformes, including American golden plovers (*kolea*), bristle-thighed curlews (*kioea*), ruddy turnstones (*'akekeke*), wandering tattlers (*'ulili*), and sanderlings (*huna kai*). Hawaiian stilts (*ae'o*) are residents, but are not marine birds. Gulls are primarily a Northern Hemisphere and continental group of birds, and few are truly oceanic. None of the world's forty-seven species of gulls breeds in Hawaii. Among the gulls, only the lava and swallow-tailed gulls in the Galápagos and the silver gull in New Caledonia breed on tropical oceanic islands. The reason that gulls have not adapted to Hawaii is unclear. They may lack the ability to exploit the available food resources or, alternatively, they may be unable to eliminate sufficient excess salt from their bodies to be able to survive in a tropical oceanic environment.

Terns are better adapted to the marine environment than gulls and are basically a tropical group. Six of the world's forty-two species of terns and noddies breed in Hawaii. Sooty terns are common throughout the Hawaiian Islands. They are among the most numerous marine birds on earth, are widespread throughout tropical oceans, and breed throughout their range. Gray-backed terns also occur throughout the Northwestern Hawaiian Islands and on Kaula and Moku Manu, but are far less common than sooties. Gray-backs are confined to the tropical Pacific Ocean, including the Phoenix, Line, Tuamotu, Fiji, and Wake islands. White terns occur throughout the Hawaiian Islands and are widespread on islands in warm waters throughout the world. Blue-gray noddies are common on the three Northwestern Hawaiian islands that have rocky cliffs or pinnacles. They are confined to the central and southern Pacific Ocean, where they nest at most island groups. Brown noddies and black noddies are common seabirds in Hawaii, to be found at virtually every island group. Both species nest throughout the tropical waters of the world.

Adaptations and Characteristics of Hawaiian Seabirds

Most of the primary adaptive trends in seabirds appeared very early in their evolution. One of the most fundamental characteristics of all seabirds is the ability to fly over the ocean. Except for penguins and flightless Galápagos cormorants, seabirds have dispersed and speciated over the oceans of the world by flying rather than swimming. Albatrosses and shearwaters use little energy when they patrol great expanses of ocean surrounding the Hawaiian Islands during the breeding season in search of localized food. Seabirds that breed in Alaska and Antarctica cope with drastic seasonal changes by migrating to more favorable locations during winter. Many Hawaiian species migrate between breeding seasons, usually to seek better feeding areas.

Albatrosses, petrels, and shearwaters are superb gliders. They use their long, pointed wings and an alternating flight pattern to rise above wave crests, then plunge down into the troughs between waves. They remain airborne by exploiting micro-differences in the strength and direction of winds just above the water's surface which are created by waves and swell. Their flight patterns describe loops, which double almost into figure 8s. Great frigatebirds' long wings and huge forked tails enable them to engage in aerial acrobatics, including mid-air stops, starts, and twists. An extraordinarily low body weight (about 1.4 kilograms) in proportion to a two-meter wingspread allows frigates to soar on thermal columns that rise above the islands in the midday Hawaiian heat. Frigatebirds are so highly specialized for flight that they cannot swim or walk—their legs and feet are so reduced that they can only perch and fly. Sooty terns and brown noddies also have long, narrow wings but have a flapping mode of flight that allows them to hover, rising and falling like helicopters over schools of tunas, porpoises, and swordfish. They can descend quickly to pluck from the water small fish and squid that are fleeing from the jaws of voracious tunas. Sooty terns are especially aerial. They apparently live entirely on the wing for up to five years, from the time they fledge until they first establish breeding territories.

The plumages of most Hawaiian seabirds are relatively drab hues of blacks, whites, browns, and grays. Males and females of most species cannot be distinguished in the field, at least by humans. The bright coloration that is found on certain species functions as breeding displays and is confined to beaks, legs, and facial skin; the feathers do not change. Male great frigatebirds inflate large red pouches on their nesting grounds during courtship and establishment of their territories. Male brown boobies develop bluish casts on their faces and feet during breeding season at the same time that the females' skin becomes yellow-green. It is also difficult to distinguish most juveniles from adults by their plumages. Young sooty terns, however, at first flight are mottled with white flecks in their wings, so that their plumage more closely resembles that of winter starlings than the sharply contrasting black-and-white plumage of adult sooty terns. Red-tailed tropicbird fledglings are densely barred on their heads and upper parts, whereas adults are mostly white. Juvenile white terns have grayish-brown wings and backs, in contrast to the immaculate white plumage of adults. Roosting seabirds spend a great deal of time removing loose feathers, lice, and ticks. Ticks are common on most Hawaiian seabirds and are sometimes swallowed by preening birds, so that some of the energy that has been "stolen" by the parasites is recycled.

Seabirds have many adaptations for swimming and diving. With the exception of sooty terns, all Hawaiian seabirds have waterproof plumage. Birds maintain waterproofing by regular preening and oiling from a gland at the base of the tail. Sooty terns have poorly developed oil glands and become waterlogged within a few minutes whenever they enter the water. Great frigatebirds are sometimes said to lack waterproofing because they have trouble taking off from the surface of the ocean, but their difficulties emanate from structural inability to take flight from the sea surface, not an absence of waterproofing. Most Hawaiian seabirds have webbed feet, which propel boobies, shearwaters, and tropicbirds in their underwater pursuit of prey. Albatrosses and boobies use large webbed feet to gain leverage when taking flight from the surface of the water. The legs of tropicbirds and shearwaters have been so highly modified to permit the use of the feet in swimming that they are barely mobile ashore. Tropicbirds and boobies have evolved special features to ease the shock of their head-first plunges into the sea after fish and squid, including thick skull bones and cushions of air sacs and feathers around their necks.

Like all birds, Hawaiian seabirds have evolved bills to meet the particular mode of life of each species (Figure 9). Bills have many functions: to seize food, attack, defend, preen, build a nest, attract a mate. Their shapes and sizes, however, are determined primarily by their use as instruments to capture food. Long bills are most useful for seizing rapidly moving prey. Short ones often provide greater gripping strength. The bills of black-footed albatrosses, sooty storm-petrels, and dark-rumped petrels are hooked, so that they can tear pieces from large squids or other prey that are too large to be swallowed whole. Hooks are obstacles to high-speed pursuit of prey underwater, and birds with such bills usually feed at the very surface of the ocean. All albatrosses, shearwaters, petrels, and storm-petrels have nostrils enclosed in tubes on the upper portion of the beak, a characteristic that has earned the order the name tubenoses. Laysan albatrosses have long bills that aid in the capture of the evasive squid, which make up a large portion of their diet. Bonin petrels use their short bills to grasp midwater fishes that surface during the night.

Brown boobies and white-tailed tropicbirds use their stout, dagger-shaped bills to capture organisms that can be swallowed whole. Masked boobies have strongly serrated bills that pose dangers to biologists whose studies necessitate the capture of these feisty birds. Because boobies and tropicbirds dive from considerable heights into the sea, their bills must be streamlined to enable high-speed pursuit of large fish. Boobies' nostrils are closed externally to prevent water from being forced into their lungs on impact with the sea. Great frigatebirds have long, slender bills with terminal hooks, which are useful for harassing seabirds and snatching prey from surface waters of the ocean. The fine, sharp-pointed bills of terns and noddies can be formidable weapons against juvenile fish and squid.

Seabirds have special problems with respect to the elimination of salts from their bodies. The kidneys of birds, like those of mammals, can dispose of only limited amounts of sodium and potassium chlorides. Because seabirds ingest large quantities of salts, they have had to develop additional means to remove excess salt from their blood. One method is behavioral—some seabirds drink fresh water when it is available, and in so doing dilute concentrations of salt to maintain a proper salt balance. I have heard Laysan albatrosses gently clacking their bills while attempting to catch raindrops during the night on Laysan. Such opportunities for Hawaiian marine birds are sporadic, and sources of fresh water are often unavailable. A more useful method has been the evolution of auxiliary desalinization systems. Nasal glands of various shapes and sizes are located in depressions above the eye sockets of albatrosses, shearwaters, petrels, storm-petrels, and terns. Excess salt is conveyed by a network of blood vessels into a series of fine tubes connected with the glands. Most species passively eliminate the concentrated solution of saltwater that collects by simply allowing it to drip off the tip of the bill, but some petrels actively eject droplets into the air through their tubenoses. The glands of boobies, tropicbirds, and frigatebirds are situated inside the eye socket. These birds disperse the salt solution that trickles from a duct in the nasal cavity by vigorously shaking their heads.

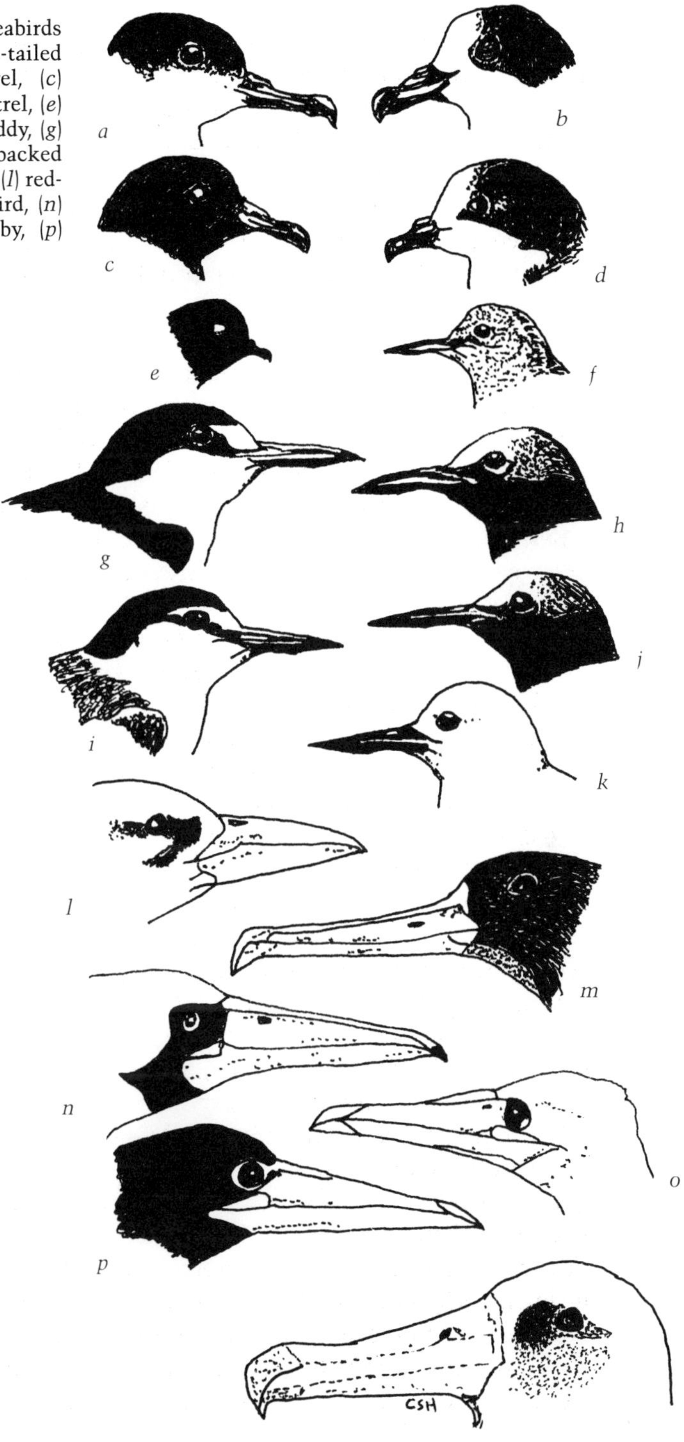

Figure 9. Bills of some Hawaiian seabirds (not entirely to scale): (*a*) wedge-tailed shearwater, (*b*) dark-rumped petrel, (*c*) Christmas shearwater, (*d*) Bonin petrel, (*e*) sooty storm-petrel, (*f*) blue-gray noddy, (*g*) sooty tern, (*h*) brown noddy, (*i*) gray-backed tern, (*j*) black noddy, (*k*) white tern, (*l*) red-tailed tropicbird, (*m*) great frigatebird, (*n*) masked booby, (*o*) red-footed booby, (*p*) brown booby, (*q*) Laysan albatross.

Hawaiian seabirds live considerably longer than land birds of similar sizes. Banding returns indicate that black-footed albatrosses live to at least twenty-eight years of age. Sooty terns, which weigh only about 200 grams, can survive to thirty-two. No doubt individual birds live longer than such records indicate because the aluminum and monel bands used by biologists to study the survival of seabirds often wear out long before the birds do. Seabirds also are known for deferred maturity—they wait until they are fairly old before they begin to reproduce. Laysan albatrosses rarely breed until they are seven or eight years old. Terns and noddies usually wait until they are at least four. Great frigatebirds first reproduce between the ages of seven and eleven. Subadults normally visit breeding colonies at least a year before they breed, and thus have an opportunity to learn courtship routines, attempt to acquire and defend a territory, and learn the locations of the best feeding areas from experienced birds. Probably each bird needs a great deal of local knowledge before it is worth attempting to raise young in the uncertain Hawaiian marine environment, where food may be scarce or difficult to locate. A characteristic related to deferred maturity is low reproductive rates. Whereas most terrestrial birds raise several young in each brood, Hawaiian seabirds raise but a single chick each year. Brown and masked boobies often lay additional eggs, but their primary function is replacement in the event that the first egg is infertile. A second chick almost never survives to fledge.

Hawaiian seabirds weigh more than tropical species elsewhere. On a global scale, birds and mammals in cold latitudes are often larger than their counterparts in the tropics. Sooty shearwaters in Alaska weigh twice as much as wedge-tailed or Christmas shearwaters in Hawaii. Albatrosses of the southern ocean are much heavier than Hawaiian albatrosses. A comparison of the ten seabird species that occur both in Hawaii and on Christmas Island, some 2,000 kilometers south of Honolulu and virtually on the equator, is illuminating. Nine of the ten species weigh more in Hawaii—only red-tailed tropicbirds are lighter. Most Hawaiian species are at least one-tenth larger, and Hawaiian masked boobies weigh more than half again as much as Christmas Island birds. The differences are not yet adequately explained, but are probably related to the fact that many Hawaiian species are at the northern limit of their breeding ranges. The somewhat cooler, subtropical waters in Hawaii may produce larger seabirds because the waters are more productive.

Hawaiian seabirds, like their counterparts elsewhere, have adapted to life at sea and on breeding islands. Evolution has shaped the anatomy and physiology of each species to allow variations in feeding and nesting habits. Despite the different origins of the three orders of Hawaiian seabirds, many solutions to common problems are similar: webbed feet, salt glands, longevity. Yet seabirds have not been uniformly successful in the Hawaiian archipelago—the populations of species differ vastly as a result of both natural and human factors.

5 POPULATIONS

Imagine standing in the midst of a swirling and squawking cloud of sooty terns on Laysan. The morning sun is still low on the horizon but already you can feel the heat of the day coming on, and the glare off the bleached white sand would scald your eyes if you lost your sunglasses. Adult terns dive and peck at your head, young chicks scurry underfoot, the cacophony of screeching and scolding is deafening. How can you count these birds? It is more pleasant but the task is no easier on Lisianski at dusk, when blizzards of Bonin petrels return to the island, their graceful movements silhouetted against a harvest moon. As darkness sets in and you become envigorated by the cool evening air, the petrels disappear into their burrows. As you traverse the colony, you suddenly lurch to one side when a leg sinks below the surface, buried in sand up to the knee. Another burrow has collapsed. How can you count these birds?

There are simply too many individuals of most species to count one by one. The best techniques that are available to estimate some populations are inherently imprecise. The blue-gray noddies and white terns that nest along the perpendicular cliff face at the north end of Nihoa are impossible to census without the use of expensive low-level aerial photographs. The technical problems in estimating populations are exacerbated by the fact that seabird populations, like those of all wild animals, are dynamic. They change over the seasons and the years. We understand some of the reasons for change, yet others remain a mystery. Nevertheless, reasonably reliable estimates of Hawaiian seabird populations do exist and general trends during the past century are known. Populations vary over time for a variety of reasons, some of which are related to the fact that certain species are more numerous than others.

Population Assessment Techniques

Seabird populations in Hawaii have been estimated at various colonies during the past century by dozens of biologists. The earliest surveys and estimates were made on an opportunistic basis. Voyagers to the Northwestern Hawaiian Islands who managed to get ashore on Nihoa, Laysan, or other colonies often made lists of seabirds, estimated their numbers, and later published their observations in diaries, reports, or scientific journals. Such estimates were rarely based on an overall plan to determine populations. Observers did not engage in rigorous scientific sampling in part because statisticians had not yet developed the necessary statistical methods. Since the 1960s, intensive survey programs have been conducted by the Smithsonian Institution, the U.S. Fish and Wildlife Service, and the Hawaii Department of Land and Natural Resources. Unfortunately, it is difficult to interpret much of the information that modern biologists have collected. Records are inconsistent as to whether estimates were of breeding birds or entire populations. Often the methods employed in surveys were inadequately described. Standardized and objective census methods were rarely used and, through no fault of the biologists, the timing and duration of visits to remote colonies were inconsistent. During the late 1970s attempts were made to improve upon these shortcomings, but the time and expense of obtaining accurate information can be so great that they are beyond the capabilities of most government-sponsored research and management programs.

Biologists now believe that for conservation and management purposes, baseline estimates of the numbers of breeding pairs of seabirds are more useful than estimates of total populations. One can estimate the number of breeding pairs of several species simply by counting active nests during the appropriate months. Nests of large and conspicuous birds, such as boobies and frigatebirds, can usually be counted fairly accurately on flat or small islands. Most other nesting birds are censused by sample plot or strip census techniques. Populations of sooty tern and Laysan albatross colonies, for example, are estimated by randomly selecting sample plots of a fixed area within the colony. All eggs and young are counted in each plot, and the number found establishes an average density of nests. The estimate of breeding birds is then calculated by multiplying the average density by the entire area encompassed by the colony. Strip censuses are similar, except that nest densities are determined within randomly chosen strips through the colony. Inaccuracies in sample plot and strip census methods emanate both from the density estimates and the measurement of the area of the colony.

Modifications to such techniques may be necessary for some species and terrains. The nest densities of red-tailed tropicbirds and brown noddies often vary among different habitats on the same island. In such cases, each habitat is estimated separately and the total population estimate is based on the summation of the density times the area of each habitat. Burrow-nesting birds, such as Bonin petrels and wedge-tailed shearwaters, present special problems and are notoriously difficult to enumerate. The number of burrows in a colony can be estimated fairly easily, but the percentage of *active* burrows must also be determined. Many are abandoned. Because some burrows are deeper than an arm's length and excavation can harm the birds, estimates of burrow use are indirect and susceptible to large errors. Rare and endangered species also present special enumeration problems. Newell's shearwaters nest in dense uluhe ferns and cannot be censused accurately without severe disturbance to the nesting area; an effort to obtain an accurate population estimate is not worth the potential damage to the birds. Similarly, dark-rumped petrels' burrows are scattered in steep mountainous areas and some nests may not be found. Obviously the breeding populations of such birds cannot be estimated accurately.

It is extremely difficult to census the nonbreeding portion of any seabird population. Nonbreeders can account for as much as 70 percent of the population of some Hawaiian seabirds and consequently play an important role in seabird ecology. Because subadults lack the responsibilities of parents, they make landfall at irregular intervals and rarely have territories. The attendance patterns of nonbreeders at colonies apparently varies with time of day and stage of the breeding cycle. Counts of roosting red-footed boobies and black noddies peak about four o'clock in the morning, and censuses at other times of the day underestimate, often grossly, the nonbreeding population of a colony. The most accurate method of enumerating the nonbreeding portion of a seabird population may be to undertake an intensive long-term study in which most birds are individually banded and identified. Such studies, however, are very time-consuming and expensive.

Recent Hawaiian Seabird Populations

About 6 million seabirds breed in the Hawaiian Islands (Tables 1–8). Including nonbreeding birds, the total population is about 15 million. The estimates in the Northwestern Hawaiian Islands (Table 1) are subject to additional imprecision because the islands are remote and because biological information is absent for some colonies. Not all species breed at the same time of the year, and researchers cannot schedule their infrequent visits to the colonies to occur at the optimal months to census them all. Nesting sooty terns, for example, can be enumerated effectively only during incubation. Once the eggs hatch, chicks form crèches and scatter when humans enter a colony, so that accurate density estimates become impossible. The difficulties in surveying remote islands have required some population estimates to be based on information collected in a single year. Because seabird populations can fluctuate annually, population estimates should be based on information obtained over many years. Many estimates are based on adjustments for nesting success and phenology which have been assumed to be constant for all colonies. The information needed to make such corrections is not available, but breeding success and chronology in Hawaii definitely can vary with year and location.

Table 1
Estimated numbers of breeding pairs, Northwestern Hawaiian Islands

	Nihoa	*Necker*	*French Frigate Shoals*	*Gardner Pinnacles*
Black-footed albatross	40–60	200–250	4,000–4,500	0
Laysan albatross	1–5	450–550	900–1,000	10–15
Bonin petrel	0	0	30–50	0
Bulwer's petrel	75,000–100,000	250–500	200–500	10–15
Wedge-tailed shearwater	30,000–40,000	1,500–2,500	1,500–1,750	25–50
Christmas shearwater	200–250	0	15–20	0
Sooty storm-petrel	2,000–3,000	*	*	0
Red-tailed tropicbird	250–300	100–150	550–600	20–25
Masked booby	250–300	250–300	500–600	125–150
Brown booby	150–200	20–25	40–60	5–10
Red-footed booby	1,500–2,000	650–750	550–600	0
Great frigatebird	3,500–4,500	700–900	350–375	0
Gray-backed tern	9,000–12,000	3,500–4,500	750–1,000	1,500–2,500
Sooty tern	10,000–25,000	12,500–25,000	60,000–78,000	250–500
Blue-gray noddy	2,000–2,500	1,000–1,500	*	*
Brown noddy	25,000–35,000	10,000–15,000	5,000–7,500	1,000–1,500
Black noddy	1,000–5,000	300–500	750–850	200–300
White tern	1,000–5,000	100–300	500–750	150–250

Note: Estimates are from Harrison et al. (1984), in which full citations can be found.
*Breeding is confirmed but available information is insufficient to support a quantitative estimate of the population.
?Breeding is suspected although no recent nests or young have been found.

Sooty terns are by far the most numerous species in Hawaii, accounting for over one-third of the breeding pairs. Laysan albatrosses, Bonin petrels, and wedge-tailed shearwaters are the next most numerous species. Harcourt's storm-petrels and dark-rumped petrels are the rarest. Sooty terns are the most common species at all islands and atolls in the Northwestern Hawaiian Islands except Nihoa and Midway. Bulwer's petrels are the most numerous species on Nihoa, and as any casual visitor to Midway can attest, Laysan albatrosses are the most abundant seabird there. Laysan and Lisianski support the largest seabird populations in the archipelago, probably because they are two of the largest Northwestern Hawaiian islands. Today all of the main islands, including offshore islets, support only about as many seabirds as Nihoa alone. Undoubtedly the main islands supported vastly greater populations before human colonization.

Historical Trends

Enough information has become available over the past century to permit us to draw some conclusions concerning recent historical population trends of most Hawaiian seabirds. Laysan and black-footed albatross populations plummeted during the early twentieth century as a result of feather hunting and

Laysan	Lisianski	Pearl and Hermes Reef	Midway	Kure
14,000–21,000	2,800–3,800	8,000–11,000	6,500–7,500	700–1,300
105,000–132,000	23,000–30,000	9,000–12,000	150,000–200,000	3,000–4,000
50,000–75,000	150,000–250,000	400–600	2,500–5,000	400–600
1,000–2,000	50–100	10	0	0
125,000–175,000	10,000–30,000	5,000–10,000	500–1,000	900–1,100
1,500–2,000	400–600	10	25–50	20–30
500–2,500	?	1,000–2,000	0	?
1,500–2,500	900–1,300	40–60	4,000–5,000	1,000–1,300
400–425	300–350	140–160	5–10	65–75
34	15–25	50–60	0	50–60
250–300	350–450	40–60	450–500	400–450
2,000–2,500	750–850	300–400	60–75	200–250
5,000–10,000	15,000–20,000	650–750	100–200	30–50
375,000–500,000	400,000–600,000	35,000–45,000	30,000–45,000	8,000–12,000
0	0	0	0	0
10,000–15,000	7,500–15,000	1,700–2,000	500–1,000	700–800
1,500–2,500	500–1,000	75–125	2,000–6,000	20–30
600–1,000	50–100	10–20	5,000–7,500	5–10

Table 2

Estimated numbers of breeding pairs, Kauai, Niihau, and vicinity

	Kauai	*Moku 'Ae'ae*	*Niihau*	*Lehua*	*Kaula*
Black-footed albatross	0	0	?	?	20–70
Laysan albatross	30–50	0	150–200	0	25–50
Bulwer's petrel	0	?	0	15–20	20–50
Dark-rumped petrel	*	0	0	0	0
Wedge-tailed shearwater	2,000–3,000	150–250	*	500–700	1,500–2,500
Christmas shearwater	0	0	0	*	75–125
Newell's shearwater	4,000–6,000	0	0	0	0
Harcourt's storm-petrel	<100	0	0	0	0
Red-tailed tropicbird	50–100	0	*	150–250	250–400
White-tailed tropicbird	500–1,000	0	?	?	*
Masked booby	0	0	?	5–10	200–400
Brown booby	0	0	*	20–40	300–500
Red-footed booby	400–600	0	?	100–150	250–350
Great frigatebird	0	0	0	5–10	250–350
Gray-backed tern	0	0	?	0	500–600
Sooty tern	0	0	0	0	35,000–50,000
Blue-gray noddy	0	0	0	0	?
Brown noddy	0	0	0	300–500	15,000–25,000
Black noddy	250–500	0	?	100–150	20–30
White tern	0	0	?	0	30–40

Note: Estimates are from various sources, 1981–1988.

*Breeding is confirmed but available information is insufficient to support a quantitative estimate of the population.

?Breeding is suspected although no recent nests or young have been found.

Table 3

Estimated numbers of breeding pairs, Oahu and vicinity

	Oahu	*Kaohikaipu*	*Manana*	*Moku Lua Islands*	*Popoi'a*	*Mokolea Rock*	*Moku Manu*	*Kapapa*	*Mokoli'i*	*Moku'auia*
Laysan albatross	*	0	0	0	0	0	*	0	0	0
Bulwer's petrel	?	*	5–10	50–75	75–150	*	5–10	5–10	?	?
Wedge-tailed shearwater	*	300–500	10,000–20,000	10,000–20,000	1,000–2,000	0	5,000	500–1,000	1,000	5,000–10,000
Christmas shearwater	0	0	0	0	0	0	40–60	0	0	0
Newell's shearwater	?	0	0	0	0	0	0	0	0	0
Red-tailed tropicbird	*	0	10–15	0	0	0	?	0	2–5	?
White-tailed tropicbird	?	0	0	0	0	0	0	0	2–5	0
Masked booby	0	0	0	0	0	0	3–4	0	0	0
Brown booby	0	0	0	0	0	0	50–75	0	0	0
Red-footed booby	400–600	0	0	0	0	0	300–600	0	0	0
Great frigatebird	0	0	0	0	0	0	0–1	0	0	0
Gray-backed tern	0	0	0	0	0	0	25–50	0	0	0
Sooty tern	0	0	60,000–90,000	0	0	0	10,000–20,000	0	0	0
Brown noddy	0	0	15,000–20,000	0	0	5–10	1,000–3,000	0	0	0
Black noddy	0	*	*	?	0	10–25	50–100	0	0	0
White tern	50–100	0	0	0	0	0	0	0	0	0

Note: Estimates are from various sources, 1981–1988. Pulemoku, Kukuihoolua, Mokualai, Kihewamoku, and Kekepa are included in the State Seabird Sanctuary but apparently have no nesting seabirds except possibly a few pairs of Bulwer's petrels.

*Breeding is confirmed but available information is insufficient to support a quantitative estimate of the population.

?Breeding is suspected although no recent nests or young have been found.

Table 4

Estimated numbers of breeding pairs, Maui and vicinity

	Maui	*Molokini*	*Moke'ehia*	*Kaemi*	*Hulu*	*Keopuka Rock*	*Moku Mana*	*Moku Hala*	*Pu'uku (Pu'uki'i)*	*Alau*
Bulwer's petrel	0	50–100	50–150	0	10–25	0	0	0	0	0
Dark-rumped petrel	400–600	0	0	0	0	0	0	0	0	0
Wedge-tailed shearwater	* †	1,000–1,500	1,000–1,500	80–120	300–400	0	?	0	1–5	*
Newell's shearwater	?	0	0	0	0	0	0	0	0	0
Harcourt's storm-petrel	?	0	0	0	0	0	0	0	0	0
Red-tailed tropicbird	?	0	0	0	0	0	0	0	0	0
White-tailed tropicbird	100–200	0	?	0	0	0	0	0	?	0
Black noddy	50–200	0	*	0	0	?	<10	20–30	*	0

Note: Estimates are from various sources, 1981–1988. The scant available information on the State Seabird Sanctuary at Papanui O Kane indicates that it has few, if any, nesting seabirds.

*Breeding is confirmed but available information is insufficient to support a quantitative estimate of the population.

†Waihe'e Point, Hakuhe'e Point, and Pauuwalu Point have small colonies.

?Breeding is suspected although no recent nests or young have been found.

Table 5

Estimated numbers of breeding pairs, Kahoolawe and vicinity

	Kahoolawe	*Pu'u Koa'e*
Bulwer's petrel	0	?
Wedge-tailed shearwater	0	100–300
Red-tailed tropicbird	*	5–10
White-tailed tropicbird	*	0
Black noddy	100–300	0

Note: Estimates are from various sources, 1981–1988.

*Breeding is confirmed but available information is insufficient to support a quantitative estimate of the population.

?Breeding is suspected although no recent nests or young have been found.

Table 6

Estimated numbers of breeding pairs, Molokai and vicinity

	Molokai	*Mokuho'oniki*	*Kanaha Rock*	*Mokapu*	*Okala*	*Huelo*	*Moku Manu*
Laysan albatross	?	0	0	0	0	0	0
Bulwer's petrel	0	10–20	0	0	0	0	0
Dark-rumped petrel	?	0	0	0	0	0	0
Wedge-tailed shearwater	0	750–1,200	80–120	0	0	0	0
Newell's shearwater	?	0	0	0	0	0	0
White-tailed tropicbird	100–500	?	0	<20	<10	<5	<5
Black noddy	100–500	0	0	*	*	*	*

Note: Estimates are from various sources, 1981–1988.

*Breeding is confirmed but available information is insufficient to support a quantitative estimate of the population.

?Breeding is suspected although no recent nests or young have been found.

Table 7

Estimated numbers of breeding pairs, Lanai and vicinity

	Lanai	*Po'opo'o*	*Ki'ei*	*Nanahoa*	*Pu'u Pehe*	*Moku Naio*
Bulwer's petrel	?	25–35	2–5	?	*	?
Dark-rumped petrel	30–50†	0	0	0	0	0
Wedge-tailed shearwater	*	50–75	6–15	?	*	?
Newell's shearwater	?	0	0	0	0	0
Red-tailed tropicbird	25–50	0	0	0	0	0
White-tailed tropicbird	50–100	0	0	0	0	0
Black noddy	25–100	0	0	0	0	0

Note: Estimates are from various sources, 1981–1988.

*Breeding is confirmed but available information is insufficient to support a quantitative estimate of the population.

†No nests have been found at Kumoa Gulch, but behavior indicates that a colony almost certainly exists on the ridge slopes at 850 meters.

?Breeding is suspected although no recent nests or young have been found.

Table 8
Estimated numbers of breeding pairs, Hawaii and vicinity

	Hawaii	*Moku Puku*	*Paokalani*	*Keaoi*
Bulwer's petrel	0	0	0	50–150
Dark-rumped petrel	*	0	0	0
Newell's shearwater	*	0	0	0
Harcourt's storm-petrel	*	0	0	0
White-tailed tropicbird	100–500	0	?	0
Black noddy	* †	?	*	0

Note: Estimates are from various sources, 1981–1988, except for Keaoi, where the population was estimated in 1946.
*Breeding is confirmed but available information is insufficient to support a quantitative estimate of the population.
†Black noddies nest in cliffs at Ka'u, along the Maniania Pali, and at Puna, along the Puna coast trail. They may nest along the cliffs near Laupahoehoe Park, North Hilo.
?Breeding is suspected although no recent nests or young have been found.

the loss of vegetation to rabbits on Laysan and Lisianski. Albatrosses have generally recovered on all islands and may even be increasing on Midway. Laysan albatross populations seem to be expanding rapidly throughout the chain. They have increased substantially on Midway during the past fifteen years, possibly because fewer naval personnel are stationed there and human-related mortality may have declined. Since 1977 Laysans have begun breeding on Kauai and since 1985 have been prospecting for nest sites on Molokai and the north shore of Oahu from Ka'ena Point to Kaneohe. They are common on the ground during winter at Dillingham Air Field and the Kaneohe Marine Corps Air Station. Laysans have recolonized the Bonins (Ogasawaras) and are attempting to colonize other locations in the North Pacific. All such observations point to an expanding population.

Because most shearwaters and petrels are extremely difficult to census, an appraisal of their historical population trends is somewhat speculative. Wedge-tailed and Christmas shearwaters suffered serious declines at the beginning of the twentieth century on Laysan and Lisianski. Wedge-tails have recovered there but Christmas shearwaters apparently have not, probably because the vegetation has not recovered to its original state. Military activities at Midway during World War II caused extensive destruction of wedge-tail burrows on Eastern Island, but such losses were balanced by habitat gains on Sand Island, where the introduction of exotic soils proved to be beneficial. Wedge-tails had been eaten to the point of elimination on Manana by the turn of the twentieth century but since then have increased to as many as 20,000 breeding pairs with the decline of human hunting pressures. Recent populations of both shearwaters have been stable in Hawaii except for declines on Midway due to rat predation. The number of Newell's shearwaters has plummeted during historical times as a result of various introduced predators and the losses of the fledglings that perish when they are drawn to the lights on buildings and other structures on their initial journeys from the mountains to the sea. During recent years the population has stabilized, if not grown, thanks to extensive federal and state recovery programs for dazed fledglings during the fall months.

Dark-rumped and Bonin petrel populations have declined precipitously on the main islands during prehistoric and historic times. As long as the colonies of the endangered dark-rumps on Maui are protected, the population will probably remain stable or increase slowly. Bonin petrels have been completely eliminated from the main islands. Their populations also suffered steep declines in the Northwestern Hawaiian Islands during the early twentieth century, when the loss of vegetation undermined soil stability at their major colonies, causing a massive loss of nesting habitat when burrows collapsed. The populations have recovered in the Northwestern Hawaiian Islands except at Midway, where rats have eliminated tens or even hundreds of thousands of breeding birds. Most Bulwer's petrel populations have probably been stable during historical times, except for losses to rats on Midway. Insufficient information is available concerning storm-petrel populations to permit even speculation about their trends.

Red-tailed tropicbirds, masked boobies, red-footed boobies, and great frigatebirds sustained losses from feather hunting and habitat destruction during the early years of this century. Except for great frigatebirds on Laysan, all populations have recovered. Great frigatebirds, like Christmas shearwaters, apparently have permanently lost nesting habitat on Laysan because of changes in vegetation wrought by rabbits. Brown booby populations on Laysan and Lisianski have probably not been affected by vegetation and have been stable during historic times. No brown boobies have nested on Midway since the 1960s, when construction activities and other human disturbances on Eastern Island destroyed their colony. Red-tailed tropicbird populations in the main islands have apparently been increasing since the 1970s.

All tern and noddy populations on Laysan and Lisianski were affected by the problems there during the early twentieth century but have now recovered. Habitat changes on Midway have both harmed and helped terns and noddies. Rats have caused the decline of gray-backed terns on Eastern. The introduction and growth of ironwood trees have provided additional nesting habitat for white terns and black noddies, allowing their populations to increase dramatically during the past few decades. Ground-nesting brown noddies have also been harmed by rats, but some birds have compensated by nesting in ironwoods, an atypical behavior in Hawaii. The black noddy population at Tern Island, French Frigate Shoals, has increased rapidly since the late 1970s, when vegetation suitable for nesting increased. Neither sooty terns nor brown noddies nested on Manana at the turn of the century. After reductions in human interference and the designation of Manana as a state wildlife sanctuary, the combined population of sooty terns and brown noddies has grown to about 100,000 pairs. White terns have also staged a comeback on Oahu and since 1961 have begun to nest in downtown Honolulu. Because blue-gray noddies are difficult to census, it is difficult to trace historical or recent population trends, but their populations probably have been stable during the past century.

Population Regulation

Why are some seabird species more numerous than others? The answer to this question must be sought in a second: How are populations of seabirds regulated? Academic biologists and ecologists have debated the latter question for decades, and although various schools of thought exist, no single answer explains all observations. Many views on the subject are based on the assumption that populations exist in some "natural" state, unaffected by humans or their activities. Proponents of population regulation theories often search for a single, global explanation that can be applied to all biological populations. Three of the major theories on how populations of species are regulated are (1) that populations are regulated by density-dependent competition for key resources, such as food and breeding sites; (2) that populations are regulated by density-independent factors, such as climate; and (3) that populations are self-regulated by social factors: the population senses what the "correct" population level should be and adjusts its reproductive output accordingly.

From the point of view of conserving and managing seabirds as a wildlife resource, global theories are usually less important than answers concerning a particular colony. How will Laysan albatross populations be affected by a certain level of squid fishing near Midway? If we want more black noddies on Lisianski, what, if anything, can be done to increase their numbers? How would seabird populations be affected by the introduction of cats or rats on Nihoa?

Although weather can be an important factor, populations of Hawaiian seabirds seem to be controlled primarily by the availability of two resources: food and nesting habitat. Which resource is most critical varies with species, location, and occasionally other factors. Whatever the situation may have been before human contact, it is certain that either food or nesting habitat can be sufficiently affected by human actions that a certain seabird population on a given island will decline. Not all human activities are negative—some improve the environment for Hawaiian seabirds. Nesting habitat for many species can be enhanced on most islands and food resources may even be improved. Ocean thermal energy conversion (OTEC), which generates electricity by bringing cold, nutrient-rich water to the surface, may enhance productivity and generally increase the availability of food for seabirds. Some floating objects serve as fish aggregating devices, and humans have probably improved foraging opportunities for the boobies that fish from buoys and piers.

The different feeding habits of seabirds provide one possible explanation for the dramatically different population levels of the twenty-two species in Hawaii. A comparison of closely related species indicates that birds that feed offshore tend to have much greater populations than species that forage fairly close to the breeding islands. Offshore-feeding sooty terns are vastly more populous than nearshore-feeding gray-backed terns; brown noddies are far more numerous than black or blue-gray noddies; red-footed boobies outnumber brown boobies. An aspect of this phenomenon may be related to the relative amounts of squid and fish that are consumed by offshore- and nearshore-feeding species. Offshore-feeding birds usually eat much larger proportions of squid than of fish. When two similar species feed offshore, as Laysan and black-footed albatrosses do, the birds that eat the most squid (Laysans) greatly outnumber their relatives. One advantage of feeding offshore is obvious. A bird that flies one hundred kilometers from Laysan to feed forages over a far greater area of ocean than a species restricted to a ten-kilometer radius. The difference between two species' feeding areas is a function of the square of their feeding ranges, so that a bird that feeds ten times as far can forage over one hundred times as much area.

The ability or tendency to migrate also apparently influences population levels. Species that leave the Hawaiian archipelago during the nonbreeding season tend to be more numerous than species that remain year round. The migratory gray-backed terns, for example, are more numerous than the resident blue-gray noddies, black noddies, or white terns. All shearwaters and petrels are migratory and vastly outnumber the resident boobies. The migratory red-tailed tropicbirds are more numerous than boobies except on islands where tropicbirds are limited by an absence of suitable nesting habitat.

The numbers of several Hawaiian seabirds are limited by nesting habitat. Birds that have been subjected to introduced predators have lost much of the otherwise suitable breeding habitat: dark-rumped petrels, Newell's shearwaters, and all ground-nesting species on Midway. Many birds on less disturbed islands also seem to be limited by the availability of nesting habitat because they compete with each other for nest sites. Wedge-tailed shearwaters that arrive on Laysan during spring evict and kill Bonin petrel chicks that have not left their burrows. Returning Bulwer's petrels treat laggard sooty storm-petrels similarly. Such species would probably be more numerous if additional space to dig nesting burrows were available. The populations of black noddies and white terns are also limited by nest sites. The planting and growth of large ironwood trees on Midway since the turn of the century has coincided with remarkable increases in the numbers of both birds because they nest in the trees. The populations may have so far increased that the availability of food near Midway *does* limit further increases; many of the younger ironwoods lack nesting white terns or black noddies. Populations of both species on Pearl and Hermes Reef, Laysan, and Lisianski seem to be limited by nesting habitat, and one means of increasing their numbers would be to plant ironwoods on atolls that lack trees. Blue-gray noddies are absent on most of the Northwestern Hawaiian islands because their preferred nesting habitat, a rocky promontory, is absent. Red-tailed tropicbirds require a minimum amount of shade from either shrubs or trees in order to nest. Their populations are limited on Manana, Kaula, Nihoa, Necker, and Pearl and Hermes Reef by the paucity of vegetation. Additional vegetation on those islands, whether achieved by natural or artificial means, would probably increase the numbers of red-tailed tropicbirds.

Food seems to be the limiting resource for many populations of Hawaiian seabirds. It is difficult to understand why populations of brown and masked boobies, which require only bare ground on which to nest, could be limited by the amount of available nesting habitat on any Northwestern Hawaiian island. For similar reasons, most islands with sooty tern and Laysan albatross colonies apparently have unused nesting space, and seabird numbers seem to be limited by the availability of food near colonies. However, an additional factor must also be considered. Most Hawaiian seabirds breed in colonies and function as social groups. Some species may not use all available nesting habit at an island for behavioral reasons. White terns and black noddies on Midway breed almost entirely on Sand rather than Eastern, although similar ironwood trees grow on both islands. Birds nesting for the first time may need the presence of other nesting birds. Little is known about the means by which seabird colonies are founded, but an absence of seabird colonies may not necessarily be due to a lack of adequate nest sites or food.

Population Variation

Hawaiian seabird populations are dynamic—they vary with season and year. The assumption that any population of wild animals is stable is usually incorrect and in any event would be very difficult to prove. On a global scale, many seabird populations are undergoing long-term population trends. Northern fulmars have been increasing in the North Atlantic for the past two centuries. Herring gulls probably number twenty times their population at the turn of the century. Many penguin species have increased dramatically in the Southern Ocean during the past hundred years, probably because whalers so decimated the whales that tremendous surpluses of krill are available for penguins to eat. On a more pessimistic note, populations of anchovy birds off the coast of Peru declined precipitously in the late 1950s and again in the mid-1960s and have yet to recover.

The number of Hawaiian seabirds ashore varies from season to season. Although some species breed over extended periods of time, many are present in the colony during certain months and absent in others. Few black-footed albatrosses are found on Laysan's beaches during September, although they are abundant in December. The numbers of birds in colonies also change with the stage of the breeding cycle. Far more black-foots are present during courtship than during incubation, when half of the breeding birds feed at sea. Once the young hatch, the parents spend decreasing amounts of time at the colony and return briefly only to feed their chicks. Similar variations occur with most species, especially with migratory birds, which leave the islands for months at a time.

Populations may vary annually because of disease and changes in oceanographic conditions. Recently the only known seabird disease in the Northwestern Hawaiian Islands has been a pox virus on Midway which has affected young red-tailed tropicbirds since 1961 and young albatrosses since 1978. When Paul Bartsch visited Midway in November 1907, however, he recorded vast numbers of dead sooty terns "as if some epidemic had invaded this colony and almost overwhelmed it." Diseases can have dramatic effects on population levels and may have broken out on remote Hawaiian islands without human detection. Seabirds in the main islands probably harbor avian malaria and other diseases derived from contact with a vast number of alien birds. Although little is known about marine poisoning of Hawaiian seabirds, seabirds have died in incidents of food-chain poisoning in Alaska and Europe. Ciguatera, a type of fish poisoning that is prevalent in Hawaii and other tropical waters, may affect Hawaiian seabirds and could account for the low population levels of nearshore-feeding species such as brown boobies. Humans who have eaten nearshore fishes that have fed on a dinoflagellate (*Gambierdiscus toxicus*) have developed symptoms of poisoning.

Oceanographic changes can affect birds through weather conditions or changes in the abundance of prey stocks. In 1964–65 the number of breeding pairs of Laysan albatrosses on Midway was half what it was in the previous three years. The decrease was temporary—many of the absent birds returned for the 1965–66 season. Boobies on Kure similarly had particularly poor success a few months later, during the spring of 1965. It may be significant that these phenomena occurred during a year of El Niño, an anomalous appearance of warm water in the eastern Pacific off the coasts of Ecuador and Peru which recurs every three or four years. This geographically restricted phenomenon is a highly visible indication that the entire eastern tropical Pacific is warming as part of related atmospheric events that are collectively called the southern oscillation. The effects off the coast of South America often cause great decreases in local seabird populations, especially species that feed on cold-water anchovies. Although the effects as far away as Hawaii are still somewhat uncertain, El Niño of 1982–83 was the largest in the twentieth century and had severe effects on seabirds at Christmas Island. In addition to changes in the food supply caused by the warm waters, hard rains brought severe flooding to Christmas Island. Total reproductive failure of all seabirds ensued and virtually all birds disappeared from the island, although most populations substantially recovered the following year. Most oceanographic changes are much more subtle than those of El Niño of 1982–83, but another source of variability in marine bird populations has been underscored.

The study of seabird populations poses many problems. They are difficult to estimate and highly variable. Populations in the Northwestern Hawaiian Islands have generally recovered from the effects of feather hunting and habitat losses during the early part of the twentieth century, but certain species on some islands apparently have not. Some populations on Midway have been reduced or eliminated by rats. On the main islands, several subspecies are in danger of extinction by introduced predators and increasing urbanization. But most species are resilient and many populations can recover even from severe environmental stress if prompt corrective actions are taken.

6 BREEDING ECOLOGY

In early November the lawns of Midway are spectacles of avian passion when Laysan albatrosses arrive in force to ensure the survival of their species. Pairs fence with beaks, rapidly shake heads from side to side, tuck bills under spread wings, bow and nod rapidly in unison. When a stranger intrudes to make the dance a threesome, a jealous male can attack viciously, driving the interloper off with shrill screams. At the height of ecstasy pairs stand on the tips of their webbed feet with heads and beaks stretched skyward, mooing loudly. The sounds of whinnies, high-pitched squeals, and bill clappering together with an orgy of mating gooneys stimulates the unpaired birds. They chase to and fro across the grass to jump on or avoid each other. Piles of four or five impassioned birds are not uncommon, and a fascinated human voyeur will rest assured that these albatrosses will replicate themselves.

All creatures must produce enough young to replace lost individuals or their populations will decline and the species will eventually become extinct. Hawaiian seabirds do so by congregating in colonies, often in phenomenal numbers, each bird tending to nest only in the vicinity of its own kind. They have evolved reproductive strategies that enable the greatest number of fit young to fledge at the least cost to adults. The conditions under which they reproduce varies with the species. Timing can be extremely important in the semiseasonal Hawaiian marine environment. Competition for nest sites on islands where suitable land is limited can be intense and results in zonation of nesting habitat. Although limitations of suitable nesting habitat may account for aggregations of nesting seabirds in Hawaii, colonies provide many advantages. Breeding success over time is the bottom line in reproductive biology, and the stages that ultimately constitute success include prelaying, incubation, and growth and development of young. An entire volume could be written on the breeding ecology of Hawaiian seabirds; only an overview is possible here.

Timing of Breeding

Breeding season for seabirds in temperate and cold waters is related to food supply. Far more fish and other prey are available to seabirds during spring and summer than in winter, when marine productivity is low. Weather can also be an important factor in determining the time of the year that is most advantageous for reproduction. Antarctic and Alaskan seabirds must fledge before autumn's frigid storms or they may perish long before reaching adulthood. Tropical seabirds on oceanic islands rarely have an urgent need to complete their nesting cycle by any particular date, and because they can breed whenever food and territories are available, they often lack well-defined breeding seasons.

In Hawaii, in contrast to many other tropical locales, the breeding season of most seabirds is predictable, if we ignore the fact that experienced breeders take occasional rest years. Year-to-year variations may change egg-laying dates by many weeks, and a few individual birds of most species may be found in any stage of the breeding cycle during much of the year, yet the reproductive patterns of most species are seasonal. Why has the Hawaiian seabird community evolved a spring-summer breeding season when other tropical colonies comprising similar species breed year round? The principal answer is probably to be found in geography. In contrast to the Seychelles, Christmas, and Ascension, Hawaii is fairly far from the equator. Many of its breeding islands are subtropical and several species are at the northern limit of their breeding range.

Many studies in temperate and cold areas have concluded that seabirds time breeding to coincide with the greatest abundance of food. In the tropical Galápagos, the availability of an adequate food supply is believed to be the primary factor that triggers the onset of breeding for shearwaters. While high latitudes often have enormous flushes of food during spring, such fluctuations in tropical oceans are generally much less extreme. However, Hawaii's subtropical location seems to give it characteristics that are intermediate between those of northern and tropical waters. Prey such as goatfishes, mackerel scad, and dolphinfishes spawn and are most abundant during spring and summer. Flyingfishes and tunas migrate north toward Hawaii from equatorial waters at the same time. As a result, Hawaiian seabirds have regular spring-summer breeding cycles because most species have evolved to lay during the months when food for their young is most available.

Another important factor in the development of a species' breeding season is the amount of daylight that is available for feeding. Although the difference in the length of a day between the summer and winter solstices is not nearly so dramatic in Hawaii as it is in temperate and northern latitudes, days are several hours longer in summer than in winter. The additional time is crucial during the chick-rearing period, when parents must forage successfully or their young will starve.

What about the five species that raise and feed their young during winter? Three of the five feed during the night and exploit different prey than the diurnal-feeding summer breeders. By breeding during winter, when nights are longer, Laysan albatrosses, Bonin petrels, and sooty storm-petrels enhance their ability to feed their growing young because they have increased time to forage. Black-footed albatrosses and black noddies also breed during winter for reasons that may be related to their feeding habits. Black-foots feed largely on flyingfish eggs, which may be most available during winter and early spring. Black noddies feed in nearshore waters, often with schools of resident predatory fishes. Thus the latter two species have developed feeding niches and breeding cycles that enable them to avoid competition with other Hawaiian species.

A third important factor in the development of a spring-summer breeding cycle is weather. Hawaii has seasons. Not only are nights longer during winter, but temperatures are cooler and storms with high winds are common. Birds that breed during winter are exposed to more severe weather conditions, which adversely affect the success rates of raising young. A biologist need not experience too many days of blowing sand on Laysan or stormy seas offshore Necker to appreciate the difficulties birds face when they breed during winter. The young of winter breeders sometimes succumb to the elements. In recent years large numbers of black-footed albatross chicks have drowned during storm tides at French Frigate Shoals. Storms and accompanying high winds have devastated black noddy colonies at French Frigate Shoals and Midway when nests were blown from exposed beach heliotrope shrubs and ironwood trees. Early-nesting sooty terns on Moku Manu drowned in heavy rains in January 1948. Wedge-tailed shearwater chicks on Manana were buried up to their necks in mud after heavy rains collapsed their burrows in October 1978. Probably food availability, day length, and weather are all important factors in the development of annual breeding cycles in Hawaii. Although it is not possible to separate the effects of the various factors, the ability to obtain food for growing young is probably the most important.

Nesting Habitat

The land available for nesting in the Northwestern Hawaiian Islands, a total of only a few thousand acres, is trivial in comparison with the North Pacific Ocean, from which millions of breeding seabirds are drawn. Although under pristine conditions the amount of nesting habitat would not have limited seabirds on the main islands, today relatively small areas are suitable for nesting, largely because of human-related habitat changes and alien species. One consequence of habitat limitations is competition among species and among individual birds of the same species for the best nest sites. Seabirds respond to competition for Hawaiian real estate by much the same means as tourists. Some engage in time-sharing by nesting at different times of the year. Such a solution is possible for birds that can successfully breed in winter and earn a living from the ocean at a time when food is generally not abundant. The majority of birds that take advantage of spring-summer feeding opportunities partition the available land vertically, essentially condominiumizing a limited resource.

Hawaiian seabirds nest in three planes: on, above, and below the ground (Figure 10). Ground-nesting birds are further distinguished as species that nest on open ground and those that require shade. Both black-footed and Laysan albatrosses nest on open ground but select sites with somewhat different characteristics. Black-foots prefer areas that are exposed to wind-blown sand, while Laysans choose sites close to vegetation. As a result, most black-foots construct their nests on exposed beaches, whereas Laysans' nests are found in the vegetated interiors of islands. Within atolls such as French Frigate Shoals and Pearl and Hermes Reef, black-foots are often the only albatrosses nesting on exposed sandspit islands. Recent attempts by Laysans to colonize the main islands indicate a preference for open areas with short grassy vegetation, such as those found on headlands near Kilauea Point, Kauai, and the golf course at Kaneohe Marine Corps Air Station, Oahu. Both albatrosses are extremely conservative in their selection of nest sites, and once a breeding pair has chosen a site, they return to it year after year.

Masked boobies, brown boobies, and gray-backed terns have no need for shade and also nest on the open ground. Masked boobies tend to nest along the perimeter of the larger islands, such as Lisianski and Laysan, but some nest inland. On the low, sandy islets of French Frigate Shoals and Pearl and Hermes Reef, masked boobies breed on the sands of the upper beaches. Brown boobies are more apt to nest inland, usually selecting sites with ground cover. On high islands such as Nihoa, brown boobies usually nest on rocky slopes or ridges, often overlooking sharp drops in elevation. Gray-backed terns are intolerant of dense vegetation and nest on the open ground near beaches or occasionally on rock ledges and ridges.

Tropicbirds, Christmas shearwaters, sooty terns, brown noddies, and blue-gray noddies nest on the ground but need shelter from the intense tropical sun. The eggs that are sometimes laid in exposed sites rarely hatch because parents are driven away by heat stress. Red-tailed tropicbirds and Christmas shearwaters nest opportunistically. Beach magnolia, beach heliotrope, and bunchgrass frequently provide suitable cover, especially toward the edges of dense cover, where few stems impede the birds' labored locomotion. At Tern Island and Midway, they nest under or adjacent to buildings. Where vegetation is sparse, rock cavities and overhangs may provide sufficient cover. White-tailed tropicbirds nest in shaded rock cavities along coastal headlands, high in the escarpment of the main islands, or on offshore stacks such as Mokoli'i (Chinaman's Hat). Sooty terns and brown noddies prefer to lay near bunchgrass or other

Figure 10. Nesting habitats of Hawaiian seabirds

vegetation, which shades chicks. Where flat terrain is unavailable, brown noddies nest on slopes or even in ironwoods. Blue-gray noddies nest only on cliffs and rocky outcrops, where holes protect them from direct sunlight.

Newell's and wedge-tailed shearwaters, all three Hawaiian petrels, and both Hawaiian storm-petrels protect their eggs and young from heat stress by nesting underground. Wedge-tailed shearwaters and Bonin petrels dig burrows several meters deep in sand or soil wherever grasses or shrubs provide sufficient stability. Where cover is adequate, shearwaters and petrels occasionally lay and incubate on open ground. On rocky islands some wedge-tails nest in depressions on cliff slopes, deep ledges, rock piles, or crevices. Newell's shearwaters burrow in steep mountainous terrain at elevations between 150 and 800 meters. Uluhe ferns are almost always associated with the burrows, and probably provide stability against soil erosion. Newell's tend to favor ridge crests or embankments. They need open downhill flight paths to become airborne, and depressions are prone to flooding in the heavy rain that falls on their mountain breeding grounds.

Bulwer's petrels select a wide variety of nest sites on both rocky and coralline islands. Unlike other petrels in Hawaii, they do not dig substantial burrows. Natural sites on rocky islands include rock rubble, crevices, caves, covered cliff ledges, and overhanging clumps of vegetation. On low coralline islands, they nest in the interstices of uplifted coral formations or in coral rubble. Bulwer's also adapt readily to sites under beach flotsam or human debris. Dark-rumped petrels nest in deep burrows or cavities beneath rocks above 2,400 meters in the main islands. They burrow among large rock outcrops and talus slopes or along lava flows where suitable underlying soil allows for the excavation of five-meter tunnels. Before they were devastated by predators and humans, dark-rumps also selected nest sites in wet forested areas below 1,400 meters. Harcourt's storm-petrels probably nest in protected crevices or rock caves in steep canyon walls. It is unlikely that sufficient soil exists in their nesting areas to allow excavation of burrows. Sooty storm-petrels use many types of nest sites, depending on local availablity. On Nihoa, they nest on inaccessible talus slides. On Laysan, they use cavities in guano piles and dig burrows into the soil under bunchgrass or beach morning-glory. At Pearl and Hermes Reef, they dig burrows in grasses or find cavities in coral rubble.

Four species nest on vegetation above the ground: great frigatebirds, red-footed boobies, black noddies, and white terns. Because their body structure makes it impossible for them to take off except from a perched position, frigates need at least a semblance of vegetation on which to construct their flimsy platforms. They prefer shrubs, and on islands where beach magnolia is available (Laysan, Lisianski, Midway, and Kure), nests are up to four meters from the ground. Where no substantial vegetation is available, frigates use low-lying plants only a few centimeters from the ground, such as nightshade (Pearl and Hermes Reef), goosefoot, ohai, and ilima (Nihoa and Necker), and even puncture vines (French Frigate Shoals). Red-footed boobies and frigates select similar nest sites. Like those of frigates, the nest platforms of red-foots vary in height from several meters to only a few centimeters.

White terns and black noddies commonly nest in ironwood, beach magnolia, and bunchgrass, but other plants may be used. Both species lay eggs on ledges of cliffs or rocky outcrops on Nihoa and Necker, where nesting options are limited. White terns lay on almost any object that provides stability, including window ledges, water faucets, fenceposts, tree crotches, and slightly elevated bare rocks.

Most Hawaiian seabirds prefer certain nesting habitats but are flexible enough to take advantage of a variety of sites, depending on local conditions. Christmas shearwaters, red-tailed tropicbirds, and brown noddies have such similar habitat preferences that they compete with one another for the choicest shaded turf at the edge of beach magnolias. Competition can be eased by many devices. The potential conflict for suitable nesting holes between Bulwer's petrels and sooty storm-petrels is partially resolved by time-sharing: Bulwer's breed in summer and sooties breed in winter. However, sooty storm-petrels that fledge too slowly are often removed from their nest holes and killed by arriving Bulwer's petrels and wedge-tailed shearwaters in late spring. Similar competition occurs between wedge-tails and Bonin petrels. Bonins nest in winter and shearwaters in summer, but Bonin chicks that have not vacated the breeding burrows before the shearwaters arrive are likely to die. Sooty and gray-backed terns also seem to compete for nest sites and the smaller gray-backs are pushed into marginal habitat.

Colonial Breeding

About 95 percent of all marine bird species breed in colonies, whereas only about one terrestrial bird in seven does so. Black noddies on Ascension prefer to nest in dense colonies where eggs fall from narrow unsuitable ledges despite the fact that wide ledges are unoccupied elsewhere on the island. Some, such as red-tailed tropicbirds, form loose groups while others, such as sooty terns, nest in dense masses. Hawaiian colonies range in size from a handful of masked boobies on Midway to millions of sooty terns on Lisianski. There are considerable advantages to congregating in one area to reproduce, particularly for novice breeders. Even where substantial amounts of suitable unoccupied ground is available nearby, most seabirds seek out nest sites near other birds; an instinct to pioneer and found a new colony seems to have limited survival value. As a result, some colonies are simply huge. The presence of a colony suggests that the area is suitable for raising young—there must be food nearby and the site is relatively safe from flooding and from exposure to high winds and predators.

Many species use visual cues from neighbors to locate food. Each morning sooty terns, wedge-tailed shearwaters, and brown noddies scatter over the ocean to forage, and when they see other birds aggregated in a feeding flock, they converge from afar to take advantage of the ephemeral food source. A colony serves as an information center so that individuals can use one another to locate patchy prey. Because the ability to find food locally may be more of a limitation on a colony's size than the amount of food available, the presence of many birds scouting for food may enhance a colony's survival.

Because ground and burrow-nesting species can be attacked by native and alien predators such as frigatebirds, rats, dogs, and humans, seabirds nest primarily where predatory mammals are few or absent. When a predator enters a colony, the warning cries of other birds, even those of another species, protect the adults. Terns and noddies sometimes mob an intruder to drive it away. A Steller's sea eagle that arrived on Kure from Asia in 1978 was mobbed by hundreds of white terns. Masked boobies charge humans who get too close to their nests, brandishing their powerful pointed bills. Black-footed albatrosses can be extremely ill tempered and will snap viciously at intruders. Both ease of food-finding and minimization of exposure to predators encourage individual birds to breed at the same time. All seabirds lay synchronously, at least to the extent that subgroups lay simultaneously. They manage to do so largely through social stimulation: the presence of a large number of mature adults synchronizes copulation and egg-laying. As a consequence, young hatch simultaneously and the colony as a whole is exposed to less risk than it would be if layings were spread out. Clustered groups of nests diminish the attractiveness of any one nest to predators. Birds nesting in the middle of a colony are protected by their neighbors on the colony's periphery, who are the first to succumb to predation. As a colony expands, a smaller proportion of its population consists of vulnerable families on the boundary.

Site fidelity is the tendency to nest in familiar physical surroundings. It is a common characteristic of many seabirds, especially albatrosses, and is most noticeable where nest sites are limited. The attachment to a nest site is proportional to the effort that is involved in acquiring it. Bonin petrels and wedge-tailed shearwaters, having expended considerable effort in digging out their burrows, tend to return to them in succeeding years. Boobies typically defend the same site each year, although masked boobies on Kure sometimes wander twenty meters or so to reestablish a territory. Great frigatebirds invest little time or energy in defending a breeding territory and establish a new one for virtually each breeding attempt. For species such as terns that nest in unstable habitats, site fidelity is much reduced. Instead, terns nest in familiar social surroundings, a tendency called group adherence. Group adherence facilitates the rapid colonization of newly suitable habitats, such as sandy beaches that are sometimes inundated by tides. In Hawaii, the movement of a colony or subcolony en masse to a new area is most common among sooty and gray-backed terns. At French Frigate Shoals, sooties abandoned Tern Island in 1942 to war activities. They returned when the navy departed in 1945, left once more when the Coast Guard moved its LORAN station from East to Tern in 1952, and came back again in the early 1970s. Each time the population apparently shifted between Tern and East. Manana, offshore Oahu, was probably recolonized in 1947 by a portion of the sooty tern colony from nearby Moku Manu. Kure was first colonized by sooty terns in the mid-1960s by birds that had been disturbed on Midway. The gray-backed tern colonies on Midway seem to change location virtually every year, probably in response to rat predation.

Colonial breeding has its disadvantages. The proximity of nesting birds facilitates the spread of diseases such as avian pox. Colonies tend to engender a great deal of stress among birds at adjacent nest sites, which sometimes results in injuries or deaths. Colonial breeding concentrates a substantial portion of a population at one location. Placing all eggs in the same basket presents additional conservation problems because a single disturbance can create a far more serious problem to a population when birds are colonized than when they are scattered over a wide area. Short-tailed albatrosses nesting on Torishima Island, Japan, almost became extinct when their only remaining colony suffered several volcanic eruptions in the 1930s.

Reproduction

Prelaying Activities

To breed successfully, each pair should have a well-defined territory. In many species, a male returning to the island of its hatching after several years at sea searches for a suitable nest site and attempts to establish a breeding territory. Many seabirds attempt to breed during the first season in which they establish an adequate site. The site may be a burrow, a branch on a beach heliotrope bush, or a bare piece of sand, but whatever location the male bird selects will be defended by him and by his mate against challengers. White terns and black noddies peck and grapple with each other, and red-tailed tropicbirds' savage fights often draw blood or result in death. For months before laying, wedge-tailed shearwaters defend their burrows and the adjacent sand, warding off intruding birds with their powerful beaks. The territorial behavior of Laysan albatrosses is less intense, and usually is limited to lunges and snaps at intruders. They often react to intrusion by bobbing forward and downward, movements that indicate the position of their nest site. Once albatross chicks have hatched and are being reared, neighboring adult black-foots and Laysans frequently approach them to peck viciously, often creating bloody bald patches on juveniles' necks. Such behavior, being directed at progeny rather than adults, indirectly achieves the desired territorial result: if a chick dies, its parents may not breed there the following year.

Curiously, great frigatebirds exhibit very little territorial behavior before pairing. Males display in groups without defending much of a specific territory and can be so packed that they are in bodily contact. Male frigatebirds direct their displays toward females, not toward other males. Once a pair has formed, both partners defend the territory by snapping, lunging, and waving their wings against intruders and neighbors. If a male is unsuccessful in attracting a female, he changes location and tries again. Boobies defend their sites with a wide variety of territorial displays that assert ownership and control. I have seen an adult masked booby on Necker savagely pummel a frigatebird chick with its bill, causing extensive bleeding. Such incidents may be rare, because few frigatebirds nest on vegetation so low that a ground-nesting booby could attack a young bird. Given the rapacious behavior of adult frigates toward boobies, the masked booby may have achieved a measure of rough justice.

Many Hawaiian seabirds are monogamous; they mate for life and few couples "divorce." Such strong pair bonds are common among seabirds and, contrary to early behavioral theory, are also common among land birds. As a rule, seabird pairs that successfully raise young stay together and pairs that fail split up and try the next year with other mates. Fidelity among wedge-tailed shearwaters is related to successful incubation and chick rearing in prior years. Half or fewer of the pairs that fail to raise a chick remain together. Male and female albatrosses, returning to their breeding territory after summering in the North Pacific, find not only their sandspit islands in the trackless ocean but also an easy reunion with their mates. Fidelity to a mate is enhanced by fidelity to a site. Most shearwaters and boobies have enduring pair bonds, but some masked boobies seek new mates without regard to their previous breeding successes. Great frigatebirds are polygamous and seemingly remate with a previous partner only by chance. They have no site fidelity, probably because their alternate-year breeding strategy does not lend itself to territorial defense. Hawaiian red-tailed tropicbirds form fairly permanent pair bonds, but attachment to mates as well as to nest sites erodes over time.

Birds' behavior is instinctive and their programmed reponses have justifiably earned them the epithet "birdbrain." Courtship behavior attracts mates and later maintains pair bonds. Outward manifestations include a variety of socially stimulating displays, many of them bizarre. Some are rather loose, others highly ritualized. Several are associated with the acquisition and defense of a breeding territory, which for many species is essential if a male is to attract a mate. Hawaiian albatrosses are so renowned for their elaborate courtship dances that they are called gooney birds. These birds have become famous for their elaborate series of displays, postures, squawks, bows, rubbernecking, and scrapes. The dances of Laysan, black-footed, and short-tailed albatrosses are similar, but the various displays have minor differences that even humans can readily learn to recognize. Such differences usually lead to a rejection of any bird that tries to dance with an individual of another species. As Duke Ellington said, "It don't mean a thing if you ain't got that swing." Breeding behavior thus serves as an isolating mechanism to maintain each species as a separate genetic entity, but the occasional gray albatrosses that are hybrids of Laysans and blackfoots emphasize its imperfection.

Shearwaters and petrels have little dramatic behavior to compare with albatrosses' antics. One wonders how mutual billing, head preening, and nuzzling can be sufficient to establish lifelong pair bonds, but probably the long hours in subterranean burrows and high-speed dual courtship flights strengthen the attachments. Shearwaters and petrels have a honeymoon period at sea after the pair has formed, which allows the female to feed while the egg is forming. Nighttime in a shearwater or petrel colony is full of the weird screechings, pathetic wailings, and deafening growls of duetting birds, which are important both in courtship and in territorial defense. The Hawaiians named the birds *'ua'u* (dark-rumped petrel), *'ua'u kani* (wedge-tailed shearwater), *'a'o* (Newell's shearwater), and *'ou* (Bulwer's petrel) in an attempt to mimic their sounds. A field camp in a wedge-tailed shearwater colony in May is not a place for a good night's sleep. Storm-petrels' colonies are also pierced by eerie flight calls of breeding birds, but they are particularly secretive and little is known of their social behavior before egg laying.

Terns and noddies are social birds with highly ritualized displays that are most evident when they gather to breed. Sooty terns congregate in dense squawking flocks above the colony night after night before they settle down en masse to lay eggs. Like other terns, they commonly strut in stereotyped parade postures on the ground, with heads extended well forward and wings held away from the body while they prance about rapidly. Brown noddies and gray-backed terns perform similar ground displays. White terns at Midway commonly fly in pairs far above the island, then glide rapidly to sea level. Such "high flights" are also common among sooty terns and may be characteristic of all terns and noddies. All species engage in some form of courtship feeding. Courtship feeding is most common among black noddies just before they lay their eggs, an indication that it provides nutrition for the female while the egg is developing.

Boobies begin their courtship with sexual advertising, such as the changing colors on the faces and feet of brown and red-footed boobies and the skyward tilt of the bills of masked boobies. Once a ground-nesting masked or brown booby pair forms, the birds run through such a large repertoire of ritualized displays that on some tropical Pacific islands they are called gooney birds, a name properly reserved for albatrosses. Such displays are common before egg laying and at ceremonies when parents exchange incubation chores. Tropicbirds court primarily in the air. Pairs of red-tails soar and glide over prospective nest sites on atolls, squawking vociferously. White-tail pairs engage in nuptial flights in Waimea Canyon and along the Na Pali coast on Kauai, descending rapidly hundreds of meters in tandem. Male great frigatebirds gather in groups to display their inflated red gular sacs to females flying overhead. Once a frigate pair is formed, they build a nest within a few days and mate. The bond is so weak that it seldom lasts more than a single nest cycle, and males may even begin breeding with another female before the chick has fledged.

Most seabirds use their bills as tools to build nests. Albatrosses fashion a hollow in soil or sand mixed with vegetation, creating a rim that can protect the nestling from winter storms. Burrowing birds use their feet to excavate a hollow before the female lays an egg. Although the burrows of some species are bare,

Bonin petrels often pack theirs with grass and other vegetation. Black noddies, red-footed boobies, and great frigatebirds make platforms of twigs, grasses, or leaves. Frigatebirds have no qualms about stealing nest material from other birds, including neighboring frigates. Many ground-nesting species such as red-tailed tropicbirds, masked boobies, sooty terns, and brown noddies get by with only a scrape in the sand or soil. White terns are the only Hawaiian seabirds that do not build at least a semblance of a nest.

Incubation

Virtually all Hawaiian seabirds lay a single egg. Brown and masked boobies are exceptions, laying a smaller second and sometimes third egg, which function as an insurance policy in the event the first is infertile. Usually all eggs hatch, and the larger nestling commits sibling murder within a week or so by evicting the smaller. Boobies usually raise but a single chick, although two brown booby pairs successfully raised two young each on Laysan in 1979. The size of all marine birds' clutches tends to be small and is related to the ability to raise young successfully. Seabirds thus fall into a broad reproductive strategy, called a *k*-strategy: they have low rates of population increase and small numbers of offspring, and breed repeatedly during a lifetime. Many fishes, in contrast, are *r*-strategists, spawning prodigiously at an early age but sometimes doing so only once or twice in a lifetime.

Many Hawaiian seabird eggs represent a large proportion of the body mass of the female, another reason for a single-egg clutch. On a global scale, birds that forage far from their colonies or in impoverished feeding areas have the largest eggs. Because a minimum size of egg is necessary to sustain the life of an embryo, the percentage of a female's body that an egg encompasses varies with the weight of the bird. Larger birds such as boobies have eggs that represent less than 8 percent of the female's weight. Eggs of the smaller terns, noddies, and petrels represent generally about one-fifth of the female's body weight, and can reach over one-quarter for diminutive blue-gray noddies. Shearwater eggs represent an intermediate proportion of the female's weight.

Unlike most terrestrial pairs, male and female seabirds take turns incubating. The growing embryo needs controlled temperature and humidity and occasional turning. In Hawaii, the summer heat absorbed by white coral sand can make the problem one of cooling rather than heating the egg, but warming can be important during winter and cool evening hours. Even after hatching, most species provide warmth or shade until the hatchlings can control their body temperatures. This ability can take several weeks to develop in ground-nesting birds, but usually only a few days in burrow-nesters. Seabirds can tolerate fairly high temperatures, an obvious adaptation to the tropical environment. Just before an egg is laid, most adults develop incubation patches of bare skin which facilitate heat transfer. Boobies lack such patches and instead incubate the eggs either under or on top of their feet.

Tropical seabirds incubate much longer than their cool-weather counterparts. Albatrosses incubate about nine weeks. Boobies, tropicbirds, and storm-petrels take about six weeks and most terns and noddies require five weeks. Causey Whittow's studies reveal that prolonged incubation requires eggshells with fewer pores than those of temperate species, in part to decrease the amount of water an embryo loses during the longer incubation period. The length of time each bird sits on the egg varies dramatically by species. Laysan albatrosses get the award for patience: each parent incubates the egg an average of three weeks before being relieved by its mate. Shearwaters, petrels, and tropicbirds typically incubate in stints of seven to ten day days. Obviously adults of such species have developed adaptations to withstand long periods without food. Boobies and most terns usually incubate for either a full day or twelve hours, but sooty terns have five-day incubation periods.

Growth and Development

Though some first-time breeders abandon their eggs, most Hawaiian seabirds have fairly high rates of hatching success. Of course, each species' hatching success varies widely by year and location. In typical years, boobies, tropicbirds, albatrosses, shearwaters, and most terns hatch about three-quarters of their eggs. Frigatebirds, petrels, storm-petrels, and white terns tend to hatch only about half. The lower rates seem to be related to disturbance at nest sites or precarious nesting habits. If an egg is lost or is infertile, many species will simply lay another one. Probably all Hawaiian terns, boobies, tropicbirds, and frigatebirds can lay twice in a season if an egg or chick is lost. The time required to replace an egg varies from two weeks to about two months, but replacement eggs tend to be laid only during the early portion of the nesting season. In contrast, no albatrosses, shearwaters, petrels, or storm-petrels in Hawaii replace lost eggs.

When a chick punctures the air space of its egg and begins calling, the parents shorten their absences from the nest in order to feed the hatchling. Most growing chicks are fed by regurgitation at least once each day, but parent Bonin petrels return to the burrow to feed their young only on alternate nights. White terns are unique in bringing fish and squid to their young crosswise in their bills. The energy requirements of most species probably peak during the first few weeks after hatching, at which time one of the parents broods the chick while the other forages. Once a young bird can regulate its own body temperature, both parents can hunt for food simultaneously.

Tropical seabird chicks are fed less frequently than their relatives in cooler climates, and as a result they grow much more slowly. Black terns in North America are fed as often as fifteen times an hour and can fledge in a mere twenty days; Hawaiian sooty terns are fed only once every sixteen hours and may take as long as sixty days to fledge. Laysan albatross chicks require almost six months. Three to four months is typical for shearwaters and boobies and six to eight weeks for terns. Slow growth is another adaptation to impoverished and uncertain food resources, and tropical chicks have much greater resistance to starvation than their temperate counterparts. Laysan and black-footed albatrosses can withstand five to six weeks without food. Albatrosses, shearwaters, petrels, and storm-petrels feed their young stomach oil, a concentrated food that aids their resistance. In a feast-or-famine environment, tropical seabird chicks must take advantage of feasts by consuming fish and squid that are much larger than they are. I have seen white tern and red-tailed tropicbird young with beaks pointed toward the sky, each with a large fish tail protruding from it. As the head is slowly digested, the fish slips farther into the gullet until at last the tail disappears.

Chicks grow in spurts that are correlated with the amount of food they ingest. The ability of the adults to transport food between feeding areas and the nest site can be a limiting factor in growth. Rarely do all parts of a nestling's body grow at the same rate. Boobies' wings continue to grow at a steady rate regardless of food shortages while bills remain the same size; thus young boobies can fledge rapidly and begin to supplement the food brought by their parents. Boobies and frigates are naked when they hatch, but all other species are covered with down. As nestlings are transformed from small balls of down to feathered creatures that resemble their parents, they begin to test their wings and develop their flight muscles. The chick's energy requirements become intense during this period. Albatrosses, petrels, and shearwaters feed their young so much that they may weigh half again as much as an adult. Later chicks are fed less frequently and must live off their body fat while they learn to fly. Some species continue to feed their young after the chicks can fly, thus assisting fledglings through the most difficult phase of their lives. Terns and boobies are usually fed for at least a few weeks after the first flight, and frigatebirds may be fed for a year or more. Although most postfledging feeding probably takes place at the colony, sooty terns and red-tailed tropicbirds may feed their young at sea.

Breeding Success

The productivity of Hawaiian seabirds varies widely by year, colony, subcolony, and species. Instances of major natural seabird reproductive failures over wide regions are well known and may have occurred in Hawaii. Black-legged kittiwakes in the North Bering Sea often have widespread nesting failures, and all seabirds on Christmas Island suffered virtually total reproductive losses during El Niño of 1982–83. Entire colonies of sooty terns and Laysan albatrosses have failed in Hawaii. Seabirds have evolved to withstand abysmal years, and an occasional bumper crop of young may be a more important attribute of an enduring species than the vicissitudes of successes and failures in individual years.

Hawaiian seabirds, especially on undisturbed Northwestern Hawaiian islands, are generally more productive that tropical seabirds elsewhere. One reason may be the high productivity of Hawaiian waters in comparison with truly tropical locations. Another may be a lack of reproductive studies on other undisturbed tropical islands. Albatrosses, shearwaters, and petrels tend to have the highest success rates among Hawaiian seabirds. Often two-thirds or more of the eggs produce fledglings, yet uncontrolled rats on Midway bring success rates at some colonies of Bonin petrels down to one in ten. Among pelecaniforms, reproductive rates are highly variable. Hawaiian red-footed and brown boobies are usually as successful as albatrosses, but only about half of tropicbird and masked booby eggs usually result in fledged young, even when sibling murders are not taken into account. Great frigatebirds are successful in only about one-third of their nesting attempts. Terns and noddies usually transform about half of their eggs to fully developed young birds, but white tern rates are closer to one-third.

Most losses occur either before the chicks are hatched or soon afterward. Egg infertility and abandonment of nests by parents are the most common causes of mortality in all species. Such problems are most likely to afflict young, inexperienced parents. Starvation and exposure to the elements are other common problems for all seabirds. Large numbers of chicks of any species can starve when local food supplies fail. More typically, starvation is an isolated occurrence when individual parents cannot locate sufficient food to feed their young, especially after a parent has died. There are no known instances of simultaneous starvation of most or all species in Hawaii. Given the wide variety of food sources and feeding methods among the twenty-two species, it is unlikely that so many different prey sources could fail at once without dramatic changes in climactic and oceanographic conditions, such as the rains that deluged Christmas Island during El Niño of 1982–83. Weather problems vary by species. Black-footed albatrosses and gray-backed terns nest near the surf zone, and can lose many nests during storm tides. High winds blow black noddy nests and white tern chicks from shrubs or trees, and virtually all such chicks perish. Storms periodically devastate sooty tern and wedge-tailed shearwater colonies offshore windward Oahu.

Rats are a common source of reproductive failure. They take eggs or young from terns, noddies, shearwaters, petrels, and red-footed boobies. Polynesian rats on Kure even attack incubating Laysan albatrosses. Larger introduced predators such as dogs, pigs, mongooses, and cats diminish breeding success on the main islands, especially among dark-rumped petrels and Newell's shearwaters. Female and subadult great frigatebirds consume many young sooty terns, gray-backed terns, and brown noddies, but probably are physically unable to reach white tern or black noddy chicks in shrubs or trees. Nihoa and Laysan finches devour any egg that is left unattended, and are a problem for terns and some petrels when humans enter a colony and disturb adults from nest sites. About one fledgling albatross in ten is devoured by a tiger shark before it gets past the reef. Sharks are attracted by this food resource and move inshore to patrol the nearshore waters of the Northwestern Hawaiian Islands during June and July.

Some reproductive losses stem from territorial interactions, especially when great frigatebirds dislodge eggs and chicks from nest platforms as they pilfer nesting material. Such losses are exacerbated when humans intrude in colonies, flushing parents from eggs and young and so encouraging neighboring frigates to raid nests. The behavior of frigates toward each other is the primary reason that they have the lowest breeding success rate of any Hawaiian seabird. Fledging rates increase in areas where nest density is low and interactions among frigates decrease. Adult tropicbirds and sooty terns sometimes kill chicks that wander into their territories, and competition for nest sites results in the eviction and death of late-fledging sooty storm-petrel and Bonin petrel chicks.

7 FEEDING ECOLOGY

The forage fish frantically try to elude the large billfish, skipjack, and yellowfin tunas. Gathering in thronging legions to feed on countless tiny plants and animals that are seasonally available, the balls of small goatfish and mackerel scad wheel and dart in the currents. They are easy pickings for the vast undersea convoys of migratory fish that venture north into Hawaiian waters during spring and summer to devour them. Silvery bodies erupt from the surface in wave after wave, then strike the water like the burst of a sudden squall. Here and there the water boils where pursuers slash through the school. Like exocet missiles, flyingfish and halfbeaks propel themselves on outstretched pectoral fins in all directions, skimming just above the surface. But escape is not so easy. The air is filled with fluttering wings and querulous cries as interlopers enter the piscine world. Plunging from above, snatching up bait fish by the thousands, seabirds converge to join in the carnage. Natural history writers long ago remarked on similar phenomena. Thomas Pennant's *Arctic Zoology* in 1785 described seabirds "watching the motions of the flying fish, which they catch when these miserable beings spring out of their element to shun the jaws of Coryphenes (dolphinfish)."

Seabirds are genuine marine creatures. They spend most of their lives at sea, earning a living from the ocean's resources. In many ways Hawaiian seabirds are truly "flying fish" because their feeding ecology is remarkably similar to that of all organisms that inhabit the epipelagic or surface waters of the sea. Food supplies, like the availability of breeding grounds, can place upper limits on seabird populations. Competition within the Hawaiian seabird community has resulted in a partitioning of food resources, and a comparison of seabird diets indicates that each species has developed its own individual food spectrum.

Although some overlap is inevitable in such a simple ecosystem, diets differ by prey species, proportions of common prey items, and size of prey. Of course, each diet is strongly influenced by where, when, and how the species feeds. Tropical seabirds are extremely opportunistic and consequently their diets are far more diverse than those of their counterparts in temperate- and cold-water ecosystems. The quantities of marine organisms consumed by Hawaiian seabirds each year greatly exceed the tonnage landed by Hawaiian fishermen.

Diets and Feeding Behavior

The twenty-two seabirds that breed in Hawaii feed on a wide variety of shoaling fishes, squid, and crustaceans, apparently taking anything they can find in surface waters which they can swallow. Their diets are complex. In a study of the species that breed in the Northwestern Hawaiian Islands, Thomas S. Hida and Michael P. Seki of the National Marine Fisheries Service's Honolulu Laboratory identified eighty-six genera of fish, eight families of squid, and eleven groups of crustaceans. Because we obtained the food samples by inducing birds to regurgitate their stomach contents, samples were usually partially digested and it was impossible to identify the species of many of the half-digested fishes and squid. Hawaiian seabirds probably eat several hundred species of marine organisms.

The diversity in selection of food is underscored by the fact that no single prey species accounts for as much as half of the diet of any seabird. Juvenile goatfishes, the most commonly eaten prey of many birds, do not make up a high proportion of any diet. They account for one-seventh of the diet of brown boobies, one-sixth for wedge-tailed shearwaters, one-fifth for white terns, and between one-fourth and one-third for brown and black noddies. No doubt these fractions would drop considerably if it had been possible to identify which of the ten species of Hawaiian goatfishes were found in the food samples. The most specialized diets are those of gray-backed terns (40 percent five-horned cowfish) and black-footed albatrosses (40 percent flyingfish eggs, but the flyingfish eggs represent at least two species). Many Hawaiian seabird diets include between twenty and forty families of prey species, in sharp contrast with the diets of seabirds in Alaska, Antarctica, and Scotland, which typically include only a handful of fishes or a single crustacean. Diet diversity in Hawaii is probably related to the patchy distribution of prey in tropical waters.

The opportunism of Hawaiian seabirds' feeding strategies is strikingly seen among blue-gray noddies. These small seabirds eat some of the largest fishes, occasionally even dolphinfishes and blue marlins. Naturally they eat the smallest larval forms of such giant fishes, but the parentage of fish fry is of no concern to a hungry noddy when it patters along the sea surface in search of anything it can grasp with its beak. Many prey organisms are taken only during certain months or only offshore certain islands, probably a reflection of seasonal occurrence. Because state-of-the-art oceanographic techniques do not permit the measurement of absolute abundance of marine creatures in the water column, biologists can only infer that prey that is eaten exclusively during certain months is locally or seasonally abundant in surface waters. Much of the variation in diets among seabirds results from the proportions of several common organisms that account for most of the food consumed by the Hawaiian seabird community. Flyingfishes, flying squid, mackerel scads, juvenile goatfishes, juvenile Forster's lizardfish, and several midwater fishes provide the bulk of the Hawaiian seabirds' food. Birds readily supplement this diet whenever other prey is available. While flyingfishes and flying squid are common in the diets of tropical seabirds throughout the world, tropical seabirds elsewhere rarely eat goatfishes, lizardfishes, mackerel scads, or midwater fishes.

Hawaiian seabirds have developed the strategy of consuming a mixed portfolio of prey species because it is impossible to rely on the availability of any particular fish or squid. William Beebe's observations of a lava flow in the Galápagos reveal the opportunism of seabirds in tropical waters:

> As molten lava reached 3000°, the ocean under the cliffs was literally boiling. A sea lion flung itself in agony from the scalding immersion, five times leaping all clear, and then seen no more. Shearwaters and frigatebirds stooped through the vapor to snatch at fish floating in the gigantic cauldron, and we saw dead petrels and shearwaters that had ventured once too often to this tempting feast.

Over the millennia, the individual birds that diversified their approaches to foraging are those that survived periodic food shortages in the unpredictable Hawaiian marine environment. The birds that eat anything are the survivors of the natural selection process. Birds that have been too finicky have joined the dinosaurs.

Hawaiian seabirds can be divided into five feeding guilds, or groups of species with closely related feeding requirements. The concept of a feeding guild groups species in accordance with ecological requirements rather than taxonomy, although taxonomy does have an important influence on the feeding habits of Hawaiian seabirds. The guilds consist of (1) albatrosses, (2) Pelecaniformes, (3) tuna birds, (4) nocturnal petrels, and (5) neuston-feeding terns. Birds within each guild have similar feeding strategies and similar diets, and eat prey of similar sizes. Because Hawaiian seabirds feed so opportunistically, their feeding habits are difficult to categorize. A focus on feeding guilds emphasizes the most common prey taken and the most common feeding methods used by each species.

Albatrosses

Each of the three North Pacific albatrosses feeds by sitting on the surface of the water and seizing prey (Figure 11), often in flocks with other albatrosses but

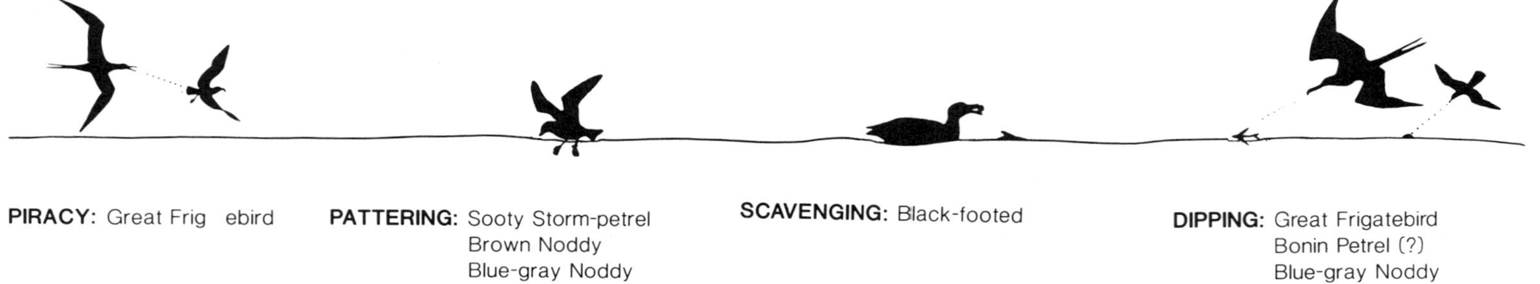

Figure 11. Feeding methods of Hawaiian seabirds

rarely with other types of seabirds. Albatrosses use their powerful bills to tear and shred large prey. Black-footed and Laysan albatrosses eat primarily squid, flyingfish eggs, and deep-water crustaceans (Figure 12). The diet of short-tailed albatrosses is less well known, but includes squid, fish, and shrimp. All Hawaiian albatrosses scavenge naturally occurring carrion or refuse from ships, but black-foots have developed this behavior into a fine art. They are not always fastidious in their selection of floating refuse. Stomachs contain such late-twentieth-century flotsam as plastic chips, rubber, styrofoam, sponges, nylon fishing line, and paper wrappers.

Eight squid families are eaten by albatrosses. Surface-dwelling flying squid are by far the most common, but most squid recovered are too well digested to be identified. When biologists develop better techniques to identify semi-digested squid, it will be possible to learn whether Laysan as and black-footed albatrosses consume different species. The proportions of fish and squid consumed by the two albatrosses are quite different. Laysans eat twice as much squid as black-foots, which eat eleven times as many flyingfish eggs as Laysans. Such differences probably result from their feeding times: Laysans tend to feed at night and black-foots during daylight hours. Both species forage in the waters north of the Northwestern Hawaiian Islands, much farther offshore than other Hawaiian seabirds. Their cool-water feeding locations imply that Hawaiian albatrosses are actually temperate species that have a somewhat tenuous relationship with the tropical marine environment.

Waved albatrosses in the Galápagos are the only other tropical albatrosses. Like Hawaiian birds, waved albatrosses eat primarily squid, fish, and deep-water crustaceans. However, they eat different squid families and do not eat flyingfish eggs. Albatrosses' feeding strategies in Hawaii and the Galápagos are similar, so dietary differences are probably due to local differences in prey species.

Pelecaniformes

The birds that make up the guild of Pelecaniformes—three boobies, two tropicbirds, and the great frigatebird—form a convenient ecological unit apart from their close taxonomic relationship. Boobies and tropicbirds plunge-dive to pursue underwater fish and squid (Figure 11) to depths of several meters. Great frigatebirds are restricted to snatching prey no more than a beak's length beneath the ocean's surface because of their structural inability to take flight from the water if they land. At sea, pelecaniforms rarely associate with one another, but frigatebirds and masked and red-footed boobies sometimes feed in flocks with sooty terns and wedge-tailed shearwaters, especially off the coasts of Central America. Brown boobies and tropicbirds are strictly solitary feeders. Most of these birds feed in deep water, but brown boobies forage inshore, often just beyond the shoreline. Red-footed boobies are especially pelagic and are found 100 to 150 kilometers from their breeding colonies.

These birds consume much larger prey than other Hawaiian seabirds. Adult

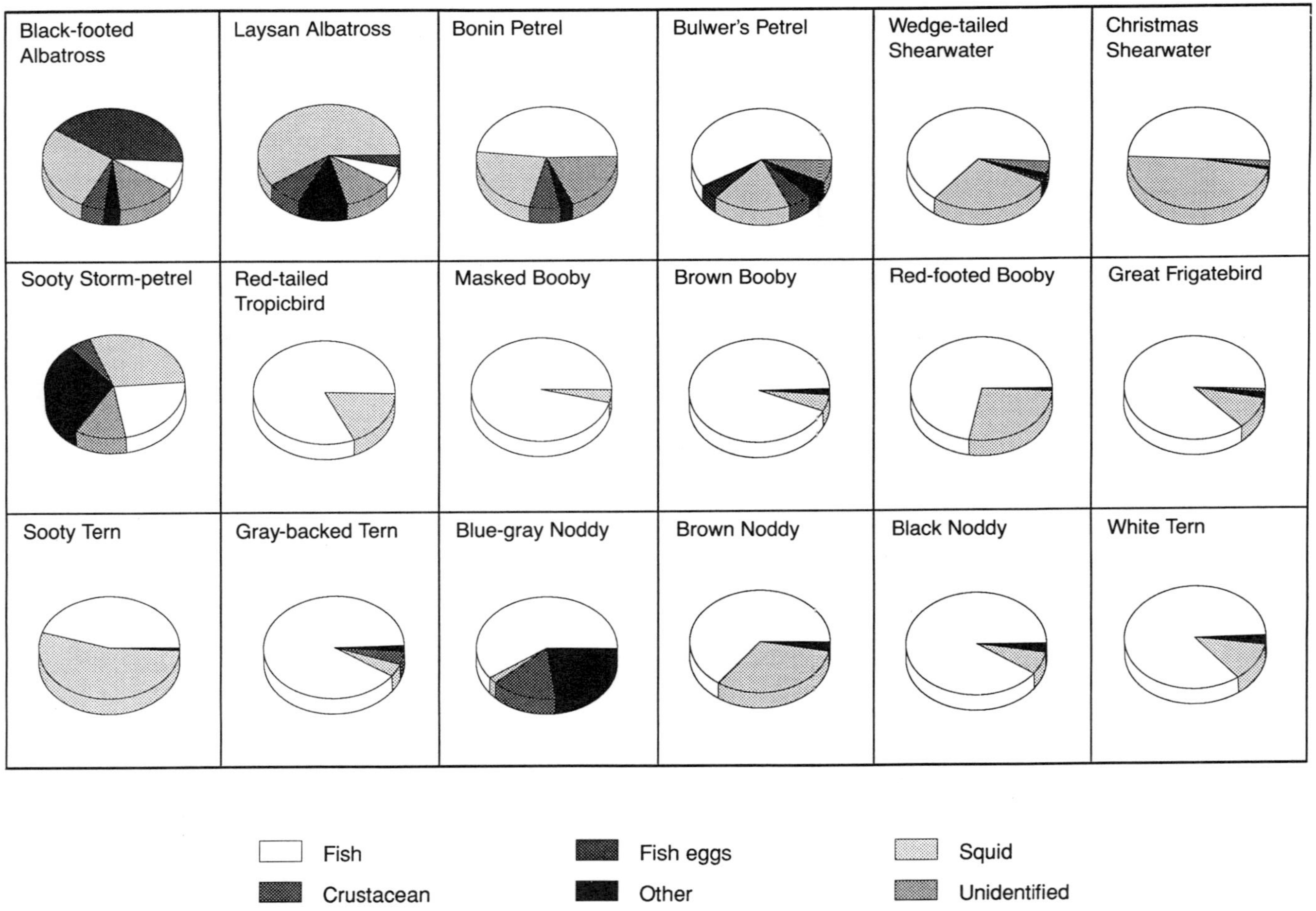

Figure 12. Diets of Hawaiian seabirds

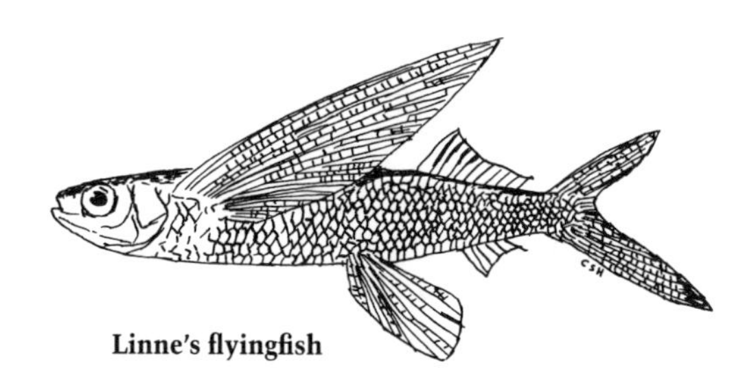
Linne's flyingfish

flyingfishes (especially Linne's flyingfish and *Cypselurus* spp.) are the most common prey, but adult mackerel scad, adult halfbeaks, and juvenile flying squid also account for a substantial portion of their diets (Figures 12 and 13). The proportions and sizes of the major prey items differ substantially among the pelecaniforms. Masked boobies take the largest, including commercially valuable fish longer than twenty centimeters. Other birds eat prey between eight and fifteen centimeters, although brown boobies eat many juvenile goatfishes that are much smaller. Red-footed boobies eat much more squid than brown boobies do—an important difference between all pelagic and nearshore feeding species. As might be predicted by their surface feeding habits, flyingfishes are especially common in the diet of great frigatebirds. Some prey species are taken seasonally or exclusively at a single atoll. Red-footed boobies and red-tailed tropicbirds eat large amounts of Pacific sauries at Midway and Kure during winter, when sauries move south with the cool North Pacific water masses. Red-tailed tropicbirds take many truncated sunfish, an open-ocean fish that is rarely taken by other seabirds, during summer at French Frigate Shoals.

The diets of Hawaiian pelecaniforms are broadly similar to the diets of boobies, tropicbirds, and frigatebirds at Ascension, Christmas, the Seychelles, Rose Atoll, and the Galápagos, where flyingfishes and flying squid are also common components of the diet. Brown boobies everywhere eat a wide variety of inshore and reef fishes. The diets of Hawaiian pelecaniforms are distinguished by the prominence of juvenile goatfishes and adult mackerel scad, which are seldom eaten elsewhere.

Tuna Birds

Tuna birds form a large and complex foraging guild in Hawaii. It comprises seven species, including some of the world's most common tropical seabirds: sooty terns, wedge-tailed shearwaters, Christmas shearwaters, Newell's shearwaters, brown noddies, black noddies, and white terns. Tuna birds forage in large flocks over feeding schools of tunas, dolphinfish, porpoises, whales, and other large predators that drive smaller prey organisms to the surface, thus making them available to seabirds. Most feeding flocks in Hawaiian waters are associated with schools of skipjack tunas. David Au and Robert Pitman have learned that flocks in the eastern tropical Pacific feed with groups of yellowfin

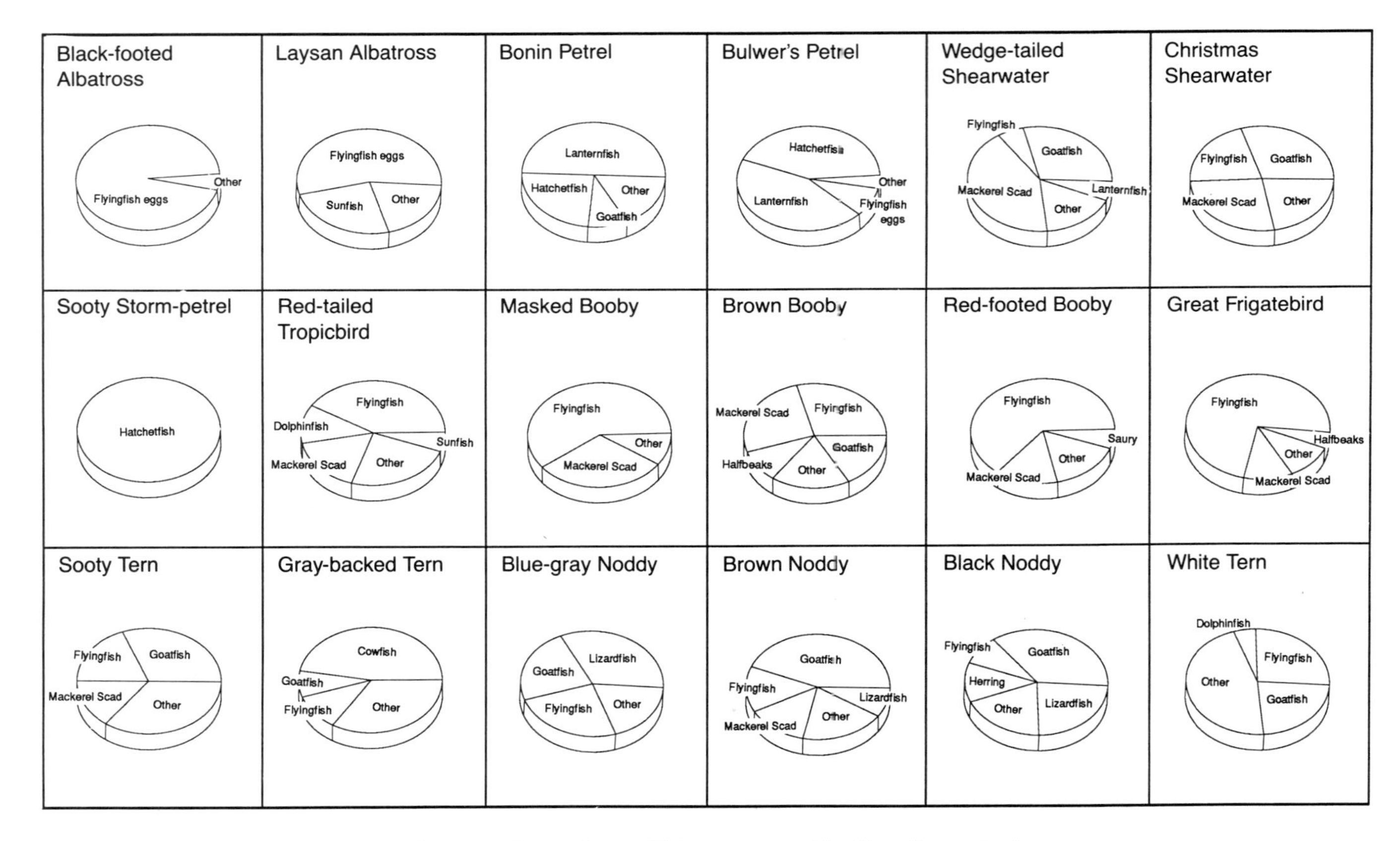

Figure 13. Proportions of fishes consumed by Hawaiian seabirds

tunas combined with spotted and spinner dolphins. Tuna birds rarely feed in the absence of fish or mammal schools and may depend on them to survive. Within a feeding flock, tuna birds use a wide variety of feeding methods to secure their prey, both on and beneath the water's surface (Figure 11). For centuries, probably as far back as the earliest Polynesian immigrants, Hawaiian fishermen have used the distinctive feeding patterns of seabirds as clues to the identity of schools of fish. When seabirds are active low over the water, they usually are feeding over skipjacks. If the flock alternates between low and high altitudes, they are probably following the deeper-foraging yellowfins. Wedge-tailed shearwaters and brown noddies fly near the surface of the water, while white and sooty terns fly much higher. As periods between surfacing fish schools become longer, the flock disperses and sooty terns fly higher and higher to act as the eyes of the flock. From distances up to eight kilometers, fishermen recognize the flash of white when sooties bank en masse from horizontal flight to make swooping dives on surfacing fish.

Tuna birds differ in their capabilities to exploit food far offshore. Sooty terns and shearwaters can forage farthest offshore, their range eclipsing that of the brown noddies. Black noddies and white terns usually feed inshore, yet some white terns are regularly seen far out at sea. All species feed on similar prey items of similar size, which are substantially smaller than prey taken by pelecaniforms. Juvenile forms of goatfishes, flying squid, mackerel scad, and flyingfishes are the primary prey consumed (Figures 12 and 13). Black noddies and to a lesser extent white terns forage with jacks and nearshore little tunas. Feeding flocks are sometimes seen hovering over jacks within a few meters of the shoreline. Inshore-feeding white terns and black noddies eat more herrings and juvenile Forster's lizardfish and somewhat fewer mackerel scads and flying squid than the others in this guild. The five remaining birds feed offshore and consume between one-third and one-half flying squid. Tuna birds select prey between three and eight centimeters long. Like all Hawaiian seabirds, they are opportunistic, and their diets change with the local availability of prey. Squirrelfish are a substantial component of sooty tern diets only during summer in the northern portion of the archipelago, but rarely are eaten elsewhere. White terns and brown noddies eat anchovies only during fall at Midway. Wedge-tailed shearwaters feed largely on gobies during fall at Manana Island, Oahu.

Given the close relationship between tuna birds and predatory fishes, it is no surprise that the birds and fishes have similar diets. There is a considerable degree of overlap among the diets of sooty terns, wedge-tailed shearwaters, skipjack tunas, and dolphinfish, both in the types and the sizes of prey taken. There is less overlap between those of tuna birds and yellowfin tuna, probably because yellowfins usually forage well below the surface. In contrast to tropical seabirds at Ascension, Christmas, and the Seychelles, tuna birds in Hawaii feed on larger proportions of fish than squid. Hawaiian birds also consume fewer flyingfishes than birds elsewhere, substituting goatfishes and mackerel scad.

Nocturnal Petrels

Bonin petrels, dark-rumped petrels, Bulwer's petrels, Harcourt's storm-petrels, and sooty storm-petrels seem to feed extensively at night. Because of their nocturnal habits and the consequent difficulties of observing them, their feeding techniques are only partially known. The three true petrels probably forage by sitting on the water and seizing prey from the surface, but Bonin petrels may also feed by dipping (Figure 11). All storm-petrels feed by pattering on the surface and do not submerge their bodies. Nocturnal petrels feed offshore, usually alone but occasionally in the company of other seabirds. They feed largely on squid, hatchetfishes, and lanternfishes (Figures 12 and 13). Most of their prey possess photophores, tiny light-emitting organs that can act as beacons to hungry birds, and rise to surface waters only at night or twilight. Food samples from nocturnal petrels are often so well digested that precise identification is impossible.

Differences in the locations of their colonies and in their breeding seasons minimize competition for food among nocturnal petrels. Bulwer's petrels, dark-rumped petrels, and Harcourt's storm-petrels breed during summer, Bonin petrels and sooty storm-petrels in winter. Dark-rumps and Harcourt's are restricted to the main islands, where the other species are rare. True petrels generally take larger prey than storm-petrels and take a higher proportion of fish than squid. The food habits of this guild are poorly known at other tropical locations, but Hawaiian petrels seem to eat higher proportions of fish than petrels elsewhere.

Neuston-Feeding Terns

The feeding habits of gray-backed terns and blue-gray noddies are unlike those of any other Hawaiian seabirds. Their diets are somewhat similar and consist of small prey obtained close to the islands. Blue-gray noddies feed in pure flocks, dipping and pattering at the surface. Unlike other Hawaiian terns, they do not depend on schools of predatory fishes to drive prey to the surface. Gray-backed terns are sometimes observed feeding at sea but probably do not feed with fish schools. They feed by plunging and occasionally associate with other terns and shearwaters.

The diets of blue-gray noddies and gray-backed terns are remarkable for the absence of squid (Figure 12). The most common prey organisms include sea striders (a marine insect), crustaceans, and juvenile forms of five-horned cowfish, flyingfishes, goatfishes, and Forster's lizardfish. Their diets vary widely in the proportions of several key organisms. Gray-backed terns eat many more cowfish, while blue-gray noddies consume more sea striders and lizardfishes. Competition is avoided to some degree by differences in breeding season: blue-grays feed most of their young two months earlier in spring than gray-backs. Furthermore, their breeding ranges overlap only on Nihoa and Necker. The food habits of blue-grays have been studied on Christmas Island, where diets are

Sea strider (*Halobates sericeus*)

somewhat similar in the importance of sea striders and minute crustaceans. However, the families of fish consumed on Christmas are quite different from those eaten in Hawaii.

Hawaiian seabird diets change with both season and island, but most variation is associated with season. In the large number and variability of their prey these birds contrast sharply with seabird communities in cold-water ecosystems. Eighty percent or more of the food of the huge seabird community off Peru consists of a single fish, the Peruvian anchovy. Many Antarctic penguin colonies have a similar dependence on krill. Seabird colonies off the coasts of Namibia and South Africa depend on but three fishes: pilchard, anchovies, and horse mackerel. Seabirds in Alaska and Scotland have relatively simple diets and gear their summer breeding strategies to take advantage of the superabundance of a few species of fish and crustaceans.

Avoidance of Competition

The simplicity of the tropical marine environment results in a great deal of overlap among the diets of the twenty-two Hawaiian seabirds. Flyingfishes, flying squid, and mackerel scad figure in virtually every diet. Yet some birds specialize in prey that others ignore. Blue-gray noddies eat large numbers of sea striders, a minute insect with an indigestible exoskeleton and fairly high levels of cadmium. Gray-backed terns eat large quantities of juvenile five-horned cowfish, a peculiar spined fish that may exude poison from its skin. Each diet has unique aspects that distinguish it from those of other birds in the community. Hawaiian seabirds have developed unique niches over the millennia by employing several mechanisms to avoid competition with one another. Some means may be important only during times of food stress, when competition among the birds could become acute. During most summers, food is probably abundant, and dietary overlap increases when birds take advantage of common prey. As most species feed their chicks in spring and summer, these are probably the seasons when demand for food is at its height, with resultant depletion of prey in the waters surrounding the colonies. Adults must obtain food both for growing chicks and for themselves. They are more constrained with regard to distant feeding locations now than at any other time of the breeding cycle because growing chicks must be fed frequently.

One means for a bird to avoid competition is to feed at a different time of day than other species. Several species have developed eye modifications that enhance nocturnal vision and feed at night. They thus can exploit a completely different class of prey than daytime feeders. Hatchetfishes, lanternfishes, bristlemouths, many species of squid, and some crustaceans migrate vertically in the water column, remaining deep beneath the surface during daylight hours but rising to the surface at night. Many of these creatures possess photophores. By feeding at night, nocturnal petrels and Laysan albatrosses have the advantage of taking different prey than most other birds. The eyes of Bonins and Laysans have high levels of rhodopsin, which enhances nocturnal vision. Black-footed albatrosses lack such an adaptation. This difference in the ability to feed at night can explain the Laysans' greater dependence on squid, which are more available to birds at night, and the reliance of black-foots on flyingfish eggs and flotsam, which are taken in daylight.

Another important means of minimizing competition is to specialize in prey of a certain size. Such specializations result from variations in the size, shape, and attendant musculature of bills. While it seems natural that masked boobies would eat larger flyingfish than noddies of one-sixteenth their weight, subtle differences in prey size may be found among birds of similar sizes. Among tuna birds, the somewhat larger shearwaters take larger flyingfishes and flying squid than terns. Among the boobies, the heavier masked boobies take considerably larger fish than red-foots or browns. The larger brown noddies take longer fishes than black noddies; the larger gray-backed terns consume bigger prey than blue-gray noddies. Even when several species appear to be feeding on the same resource, they often select prey of different sizes.

Feeding area is yet another mechanism by which birds reduce competition for food. Although precise feeding locations of breeding Hawaiian seabirds are unknown and probably will remain so until there are further improvements in marine radiotelemetry, several generalizations can be made. Albatrosses feed far offshore, probably hundreds if not thousands of kilometers north in the cooler and more productive waters of the Kuroshio. By traversing such distances, they take food that is unavailable to other species. Brown boobies feed close to shore and consequently take more prey that is associated with coral reefs than red-footed boobies, which feed fairly far offshore. Among tuna birds, black noddies and white terns feed much closer to shore than brown noddies, wedge-tailed shearwaters, and sooty terns, even though each species seeks out schools of shoaling predatory fishes. Offshore-feeding tuna birds sometimes bypass feeding opportunities in nearshore waters when they commute to distant feeding grounds.

Differences in feeding behavior further divide up food resources. Some species have evolved means to catch prey that is unavailable to others. Seven Hawaiian seabirds depend heavily on fish schools to obtain prey; fifteen do not. Among tuna bird flocks, sooty terns feed by plunging to the surface, taking prey at the interface of air and water. They cannot get wet without becoming dangerously waterlogged. Because they are faster and more agile than other tuna birds, they are usually the first to arrive at a shoaling fish school and feed before the rest of the flock arrives.

Shearwaters, in contrast, dive into the water and can swim below the surface, paddling and using vigorous wingbeats to pursue prey. They take squid and fishes in water too deep for sooty terns. Pelecaniforms also feed at different depths. White-tailed tropicbirds hover ten to fifteen meters above the ocean surface, then tuck their wings close to their bodies and plunge to depths of three meters or so. Masked boobies dive even deeper. Great frigatebirds, like sooty terns, do not enter the water but are confined to plucking prey from the surface. Some frigates use their superior speed and agility to steal food. A frigate will harass a tropicbird, booby, or shearwater until the pestered bird disgorges its last meal, which the frigatebird immediately swallows. Although frigates are renowned as man-o'-war birds because of such aerial piracy, theft accounts for only a small proportion of their diet.

All feeding in the tropics, in contrast to cooler zones, is restricted to the first few meters of the sea surface. No tropical seabird forages to the depths that cormorants, penguins, and murres do. One likely explanation is that sharks and other predatory fishes pose too great a threat to diving seabirds in the clear tropical waters; no deep-diving species could survive in tropical waters.

Choice of breeding season, as we saw in chapter 6, is a means of avoiding competition for prey. Several species with similar diets breed at different times of the year, thus minimizing competition for food during the critical chick-rearing period. Two nocturnal petrels have diametrically opposed breeding seasons. Bulwer's petrels breed during summer, laying in May and June. When adult Bonin petrels arrive to concentrate their foraging activities near the colonies in winter, Bulwer's petrels have already migrated far out to sea. Though wedge-tailed and Christmas shearwaters have similar diets, their chick-feeding activities overlap only slightly. Christmas shearwater chicks are fed from June to mid-September, whereas wedge-tailed young are fed from mid-August to November. The shearwaters do not compete for food during most of the critical chick-rearing season, a strategy also employed by blue-gray noddies and gray-backed terns.

Consumption Rates: The Impact of Birds on the Marine Ecosystem

It is difficult if not impossible to measure directly the amount of food that seabirds consume. Instead, biologists must use mathematical models to estimate the energy that Hawaiian seabirds need for various aspects of their life cycles. Such a model has been devised. In simplistic terms, estimates of the amount of food that each seabird needs for daily existence were multiplied by the total number of birds on each island, including breeding and nonbreeding birds. The estimates were adjusted downward for the portions of the year during

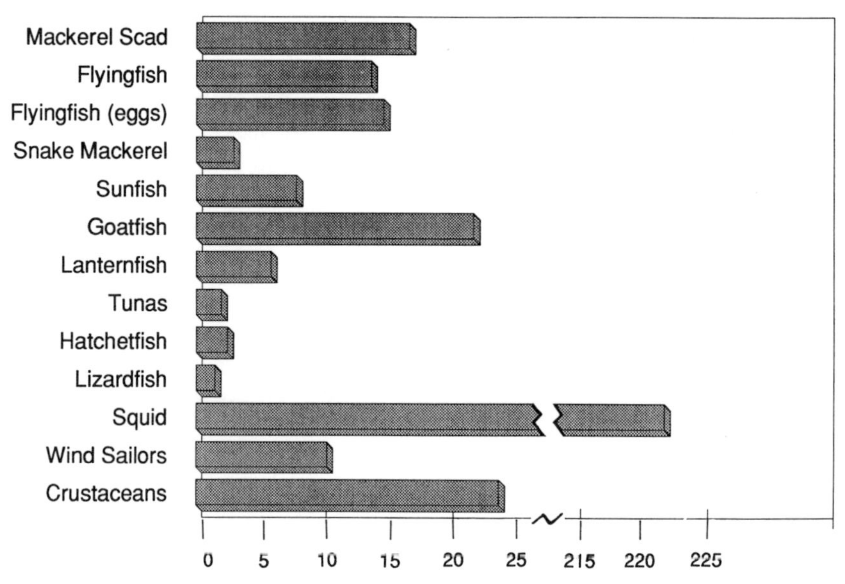

Figure 14. Estimated annual consumption of marine resources by Hawaiian seabirds (metric tons × 1,000)

which migratory birds are absent from the islands and upward for the additional food required by reproductive activities. Bomb calorimetry has been used to determine the amount of energy contained in seabirds' food sources. From this model, energy requirements were translated into an estimate of the food requirements of the Hawaiian seabird community (Figure 14).

Hawaiian seabirds consume over 400,000 metric tons of fish, squid, crustacea, and other food sources each year. By far the greater part of the consumption takes place in the Northwestern Hawaiian Islands, where 14 million of Hawaii's 15 million seabirds are found. Squid, especially flying squid, is the largest component of the community's diet and accounts for well over half of its annual consumption of prey. Fish constitutes about one-quarter of all food consumed, the bulk consisting of flyingfishes, goatfishes, and mackerel scad.

Albatrosses and tuna birds account for almost all of the prey consumed. Albatrosses eat almost two-thirds, a reflection of both their substantial populations and their relatively large size. Tuna birds take almost 30 percent of the annual consumption. Five of the twenty-two species account for 94 percent of the prey consumed: Laysan albatrosses, black-footed albatrosses, Bonin petrels, wedge-tailed shearwaters, and sooty terns. The neuston-feeding terns take trivial amounts of the resources consumed by the community, and the pelecaniforms take only about 2 percent.

A comparison of the estimates of prey consumed by Hawaiian seabirds with the present and projected fishery landings in Hawaii is interesting. Fisheries are fairly well established in the main islands, but the waters of the Northwestern

Hawaiian Islands remain relatively unexploited. Although fish landings fluctuate over the years, the 6,000 metric tons that were reported as landed by Hawaiian fishermen in 1978 represent a fairly typical annual haul. Even if fishermen report only about half of their catches, Hawaiian seabirds consume well over thirty times the annual human catch in Hawaii. The most optimistic estimate for future fishery landings in Hawaii is about 50,000 metric tons, about one-eighth of the amount taken by seabirds. Of course few of the juvenile forms of fish and squid that make up much of the diet of Hawaiian seabirds would survive to attain commercially valuable sizes. Two-thirds of the annual consumption is accounted for by squid, which are not exploited commercially in Hawaii.

An ecosystem model for French Frigate Shoals indicates that seabirds consume over two-fifths of the annual production of small surface pelagic fishes and squid. Such estimates are complicated by the imprecision of estimates of nonbreeding seabird populations, and more accurate information might change the results of the model substantially. Estimates of the effects of seabirds on marine ecosystems in cool-water systems in Oregon, Scotland, Peru, and Namibia indicate that birds take about one-fifth of the annual fish production. Seabirds are an important component of the marine ecosystem in Hawaii.

8 PELAGIC ECOLOGY: LIFE AT SEA

The salt air is fresh and invigorating as the NOAA research vessel *Townsend Cromwell* plies its way up the Northwestern Hawaiian chain from the main islands. There is usually a following sea and, with luck, the skies are clear. Life for the biologist during such days is most pleasant. Black-footed albatrosses often circle the ship for an hour or more, soaring over the gentle swells in ceaseless vigil, ready to alight and get their beaks into the latest installment of garbage that the steward tosses over the stern rail. At such times, the mind's eye conjures up visions of how sweet life must be for a seabird.

Return trips down the archipelago over the restless texture of the open sea are seldom so serene. The vessel pounds against the oncoming swells so that you can never be comfortable, even in relatively calm seas. At the worst moments during storms that seem interminable, life at sea is horrible. For days on end the deckside weather can be unfriendly. Frothing seas and harsh, gale-force winds force you to brace yourself for each step. Even in Hawaii temperatures can be cold. Eventually the ship's interior becomes claustrophobic and you lie in your bunk musing on possible weak points in the hull. At such times there is little to envy in the life of a seabird. One can appreciate the sentiments of Henry Palmer in 1891 at the conclusion of the Rothschild expedition:

> At last we reached Honolulu after a month's dreadful journey from Midway Island. I was never more pleased in my life than when I found myself on shore in Honolulu once more. So ends three months' suffering under the famous Captain F. D. Walker.

Many Hawaiian seabirds spend most of their lives avoiding the perils of the sea. Truly pelagic species, such as albatrosses, live far from land near their food supplies for much of their lives. Human contact and knowledge of these creatures is fairly well limited to the portion of their life cycle that is passed on land. Until recent years, even the distribution of common tropical seabirds in the Pacific Basin was poorly known. Distributional accounts were anecdotal and confused by the outward similarity of many seabirds that cannot be distinguished with confidence at sea except under the most fortunate of circumstances. Yet a surprising amount of information has been gleaned about the pelagic ecology of Hawaiian seabirds from recoveries of banded birds, shipboard surveys, and behavioral observations at sea.

Scores of biologists crisscrossed the tropical Pacific on vessels as part of the Pacific Ocean Biological Survey Program in the late 1960s and carefully recorded their positions and seabird observations. The gross patterns of tropical seabird movements have been sketched from the resulting distributional maps. Differences in color phases, such as the light and dark phases of wedge-tailed shearwaters, have increased our understanding. All Hawaiian wedge-tails are light-phase birds, but south of the equatorial countercurrent dark-phase birds predominate. Because of such natural markings, we know that Hawaiian wedge-tails leave the North Pacific entirely during winter, and that the dark-phase birds that appear there have migrated from colonies farther south. Many fundamental questions about the seaward activities and movements of Hawaiian seabirds will never be answered unless radiotelemetric technologies are developed to the point at which biologists can follow the movements of individual birds at the various stages in their breeding cycles.

Migration

True migration involves a clear seasonal shift from a colony to a well-defined nonbreeding area, with a return before the following breeding season. A migration is often difficult to distinguish from a dispersal or nomadic wandering. Many tropical seabirds leave the Hawaiian Islands with a strong directional bias but do not seem to move toward any clearly defined area. About half of the twenty-two breeding Hawaiian seabirds are year-round residents. The remainder are migrants, wanderers, or nomads, and leave the waters adjacent to the colonies each breeding season after their young have fledged. Some Hawaiian seabirds migrate individually, others move as distinct flocks, and in a few instances adults and fledglings travel in pairs. How seabirds find their way over the featureless ocean is still a mystery, but fledglings have an obvious advantage in learning the migration route from experienced birds.

Black-footed and Laysan albatross parents leave the Hawaiian colonies in June and July and do not return until late October. A general exodus from the breeding colonies begins in March and April, with black-foots leaving somewhat earlier than Laysans. Chicks are left to fledge on their own, and few young birds that remain on the breeding islands in August will survive to reach the sea.

These movements are part of a major latitudinal shift of all North Pacific albatrosses, whether breeders or nonbreeders, during the summer months. Albatrosses occupy the southern portions of their ranges during winter and the northern portions during summer, especially in the central Pacific. Although some Laysan albatrosses range as far south as 8 degrees north latitude in February, all movements away from the Hawaiian Islands during summer are northward. Only an occasional albatross is observed in the central Pacific in July and August and virtually none is found in September. Albatrosses can cover vast distances in a short period of time, even during adverse weather conditions.

Although both species migrate north and are generally found throughout the North Pacific during the warm summer months, detailed studies indicate that most Laysans summer in the western and central portion of the North Pacific while black-foots summer in the eastern and central portion. Laysans are fairly common in the Gulf of Alaska in May and move westward in late spring, at the same time that Laysans off California, Oregon, and Washington move seaward. Laysans apparently concentrate during summer between the western Aleutians and the Kuriles, near the Soviet Union. Although some black-foots are found off Japan and the Sea of Okhotsk in summer, they are far more common over the waters of the continental shelf off the west coast of North America, possibly an indication that they prefer somewhat warmer waters than Laysans do. For decades few Laysans were seen near the main islands during winter, but recently offshore distribution patterns have changed as Laysans have attempted to establish colonies on Kauai, Oahu, and Molokai. The numbers of black-foots dramatically increase near the main islands during the breeding season, possibly because they are attracted to the many ships that use Honolulu's port.

Shearwaters, petrels, and storm-petrels, like albatrosses, tend to depart from Hawaiian waters during their nonbreeding months. Although equatorial populations of wedge-tailed shearwaters do not migrate, Hawaiian birds winter and molt in the east-central Pacific. Few wedge-tails move either north or west from Hawaii. By mid-November, after the young of the year have fledged, both adult and young wedge-tails form large flocks offshore the breeding colonies just before they begin their migration. Although their precise routes are unknown, Hawaiian wedge-tails probably disperse south from Hawaii to the equatorial countercurrent and then east to the coast of Central America, where they feed in flocks with mixed schools of yellowfin tunas and spinner, spotted, and common dolphins. They probably return via the north equatorial current. The evacuation of the North Pacific is so complete that by February almost no Hawaiian wedge-tails are observed north of 10 degrees north latitude. Wedge-tails begin to reenter the North Pacific in March and are common within eighty kilometers of the Hawaiian Islands from April to November. Christmas and Newell's shearwaters arrive in Hawaiian waters in March, attain peak populations in May, and leave the waters adjacent to their breeding islands in fall. Christmas shearwaters probably disperse in nearby tropical waters and Newell's may wander to the north and east, but information is scanty.

The nonbreeding locations of Bonin petrels, Bulwer's petrels, and dark-rumped petrels are poorly known, but each species seems to disperse to the north and west of its Hawaiian breeding colonies, with little movement to the south or east. However, a banded dark-rump from Hawaii was collected in the Moluccas, an indication of at least some dispersal to the western Pacific. The timing of dispersal varies by species. Bulwer's and dark-rumps migrate during winter, but Bonins do so in summer. Dark-rumps are concentrated at sea just north of Oahu during the summer breeding season but may range north to the North Pacific subtropical convergence (Figure 2) from December to February, somewhat farther north than Bonins or Bulwer's. Large numbers of Bonins move north in May, toward Japan. Bonins are widespread in the waters off Sanriku and east of Honshu, but such birds did not necessarily originate in Hawaii. Their numbers at sea decline during August, when Bonins return to their colonies. The migratory habits of Harcourt's and sooty storm-petrels are very poorly understood because their Hawaiian populations are low, their pelagic ranges are extensive, and they are especially difficult to identify at sea. Sooties leave Hawaiian waters in spring and return in fall. Storm-petrels probably do not migrate long distances, and may simply scatter to the north and northwest, where they are common. Harcourt's remain south of 28 degrees north latitude, but sooties range north to the Pacific side of Honshu, Japan.

Tropicbirds are among the most solitary of seabirds, and concentrations at sea of either red-tailed or white-tailed tropicbirds are rare. Red-tails and some white-tails migrate from Hawaii during the nonbreeding season. Juvenile red-tails wander widely and may visit the eastern Pacific. Adults are frequently found in mid-ocean between 10 and 36 degrees north latitude. The numbers of red-tails in Hawaii are lowest just after the young fledge in fall and remain low from December to February. They return in March, and their numbers at sea peak in summer and fall, when populations are scattered throughout Hawaiian waters. Some white-tails disperse from the main islands to the south and west during fall, but many can be seen year round soaring in Waimea Canyon or along the Na Pali and Kaholo Pali coasts.

Hawaiian boobies and great frigatebirds tend to be sedentary, roosting on the colonies each night all year. Such habits severely constrain offshore movements. Juveniles are exceptions and frequently wander thousands of kilometers throughout the Pacific without seeking landfall. Masked and red-footed boobies banded in the Northwestern Hawaiian Islands have been recovered at Johnston, Wake, and the Marshalls. Juvenile great frigatebirds are particularly nomadic; many birds banded on Laysan and at Pearl and Hermes Reef have turned up in the Philippines. Although the numbers of boobies and frigatebirds ashore during daytime decline in the winter nonbreeding season, the birds have not migrated. Most adults remain in Hawaiian waters, but in the absence of reproductive duties spend more time at sea than ashore.

Hawaiian blue-gray noddies and black noddies are residents and do not migrate. They are sedentary and return each night to the colony to roost. Except perhaps when a storm causes them to disperse, neither species is encountered far from a colony. Brown noddies disperse from colonies during winter but do not seem to move more than a few hundred kilometers offshore. Gray-backed terns and most white terns leave Hawaiian waters during the winter nonbreeding season. The wintering range of gray-backs is unknown because they are rarely observed at sea. Although some white terns remain year round near Oahu and other breeding islands, their numbers substantially decline during winter. White terns tend to move south and west from Hawaii, but their precise wintering areas are unknown. Sooty terns are unaffected by an absence of land and venture farther than any other Hawaiian tern. During August breeding birds begin to spend less time ashore and become more ocean-oriented. They spend about a month offshore near the colonies before dispersing westward as far as the Philippine Sea. Birds banded on Hawaii have also been recovered in Japan, Guam, New Guinea, and Fiji. Substantial numbers of sooty terns remain fairly near Hawaii, especially at a rich feeding area 500 kilometers west-southwest of Oahu, but their numbers in Hawaiian waters drop sharply during fall.

Why do seabirds migrate? In cold northern areas, the drastic changes in weather provide an undeniable incentive for birds to relocate during the months of harsh climate. Although there are seasonal weather variations in Hawaii, it seems unlikely that weather is ever severe enough to account for large-scale movements of seabirds. The decisive factor is more likely a desire to improve feeding opportunities.

Seabirds that migrate take advantage of feeding grounds that are too distant from the colonies to be used during the breeding season. Migration allows them to exploit the full potential of the patchy distribution of food in the ocean, including concentrated food at distant convergences, fronts, and upwellings. Birds that migrate leave behind halos of reduced fish populations around the colonies and competition among a vast number of seabirds. Distant feeding areas may have few seabirds. From an annual perspective, migration improves feeding opportunities and should enhance the survivability of the species. The species in Hawaii that migrate tend to be the most numerous: sooty terns, albatrosses, wedge-tailed shearwaters. Migration has been a successful strategy for many Hawaiian seabirds.

Survival at Sea

A fledgling that is strong enough to reach the ocean has passed its first major obstacle. Yet surviving the first few weeks or months on land has little to do with survival at sea. Losses of young birds are typically very high during the first year, and fledglings are especially likely to perish during their first few months, while they perfect their skills at locating and catching food. Those that endure the first year at sea are almost as likely as adults to live through subsequent years.

One means to increase a young bird's chances of survival is for adults to care for it when it first goes to sea. Such behavior is common among many northern seabirds, including murres, puffins, and murrelets. Few Hawaiian seabirds, however, accompany fledglings away from the colony. Albatrosses, petrels, shearwaters, and storm-petrels apparently never do so, but instead fatten their young at the colony before abandoning them. Fledglings must learn to fly and survive at sea entirely on their own. Hawaiian terns, tropicbirds, and boobies often continue to feed fledglings at the colony even a month or more after the first flight. Female great frigatebirds return to the nest to feed begging young for up to a year after they have fledged. But feeding and care at sea are far less common. Adult-juvenile pairs of sooty terns are often observed far at sea vocalizing to each other. Despite problems associated with waterlogging, Patrick J. Gould has observed during fall a few adults feeding juvenile sooties on the water, hundreds of kilometers from land. Pairs of adult and young red-tailed tropicbirds sometimes are seen five hundred kilometers from land, circling research vessels and calling to each other. No one has seen a parent tropicbird feed a juvenile at sea, but parents may enhance the survival of their offspring by guiding them to the best feeding areas and providing examples of how to feed during the crucial first few months at sea.

Location of food is no doubt the greatest obstacle to fledglings when they first encounter the watery medium that will be their home for most of their lives, but other problems may be more immediate. Many seabirds enter the sea when they are still learning to fly and cannot properly take off or land. For birds that sit on the water, the constant motion of the waves is a new experience after the halcyon days at the nest, where the young bird simply waited for a parent to arrive with a meal. And hazards in the sea are legion. Tiger sharks move inshore from the outer fringes of the reef in June and July to patrol the lagoons surrounding the breeding islands. Young albatrosses entering the water for the first time have been fattened up to ensure against starvation during the first lean weeks at sea. Many unwary gooney birds never get beyond the reef, falling prey to tiger sharks, whose stomachs may contain as many as thirteen young albatrosses. Galápagos, ocean white-tipped, and gray reef sharks undoubtedly take their toll as well. The problems do not stop at the reef. The stomachs of tunas, marlins, and dolphinfish sometimes contain terns and petrels. No doubt the salmon and blue sharks that ply cooler northern waters will strike at and seize almost any bird, adult or juvenile, that sits on the surface of the ocean. One-legged albatrosses are common in Hawaiian colonies, vivid reminders of last-minute escapes from the jaws of death.

Weather can be a major threat to seabirds. They move both intentionally and unintentionally in response to weather patterns, and the 150-kilometer winds that can be associated with typhoons and cyclones often carry birds far beyond their normal ranges. During storms, petrels and shearwaters sailing in the troughs are dwarfed by the crests of waves towering far above. It seems a miracle that any are not overpowered. Young birds and adults in molt are most vulnerable to storms, when birds are battered by the elements and feeding is probably impossible. It seems likely that prey submerges out of reach and decreased visibility complicates the task of finding tuna schools. "Wrecks" where thousands of emaciated seabirds are blown ashore after severe storms are known around the globe. The gales and heavy rains associated with El Niño in 1982–83 may have resulted in wholesale starvation of certain seabird populations, but it is difficult to learn of wrecks in the middle of an ocean far from islands or observers.

With the exception of great frigatebirds and sooty terns, Hawaiian seabirds spend much of their time at sea resting on the water, especially after feeding. They are frequently encountered in huge flocks—some 4,000 birds stretching over five kilometers were observed 300 kilometers south of Oahu in May 1966 and another large milling mass of 3,500 birds has been seen near Kaula. Birds at sea concentrate near flotsam, especially patches of algae, which often are found near fronts and upwelling areas. As fishermen have learned by the use of floating aggregating devices, flotsam attracts juvenile flyingfish, mackerel scad, and zooplankton. Concentrations of marine life and enhanced local productivity have not escaped the attention of opportunistic terns, boobies, tropicbirds, and shearwaters, which can be as much as forty times as dense near flotsam as they are elsewhere. One explanation for the recent increases in the numbers of Laysan albatrosses is the rich community of marine life that is attracted to floating plastic, which provides a new source of food for Laysans. Wood, styrofoam, and sea turtles provide resting sites for terns and boobies, which feed nearby.

Hawaiian seabirds spend much of their lives winging over blue seas. Much of their behavior, habits, and movements in this domain is still poorly known. It is relatively easy to learn of their breeding habits in their island colonies; it will take a great deal of time and effort to unlock the secrets of their lives at sea.

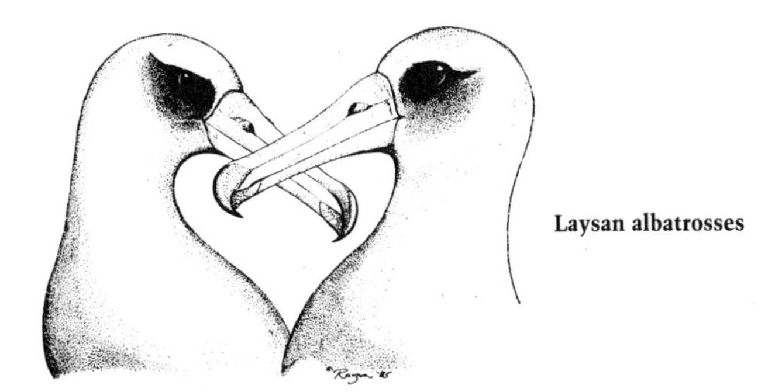

Laysan albatrosses

PART III

Hawaiian Seabirds: Family Groups and Species

Against the illimitable blue of the sky, over the unfathomable blue of the ocean the seabirds of the Pacific wing the cycle of their lives. For them the ocean is a larder: the islands and atolls their mating ground and nurseries.

—George C. Munro, 1944

9 ALBATROSSES
Family Diomedeidae

A Laysan albatross in flight is a symphony of fluid, graceful motion. During a North Pacific storm, a sailor cannot help respecting a creature that displays such mastery of the sea during its wildest moods. An albatross can glide calmly with barely a wingbeat, rise with ease over the crests of the highest waves, then swoop gracefully into the valleys between the swells. The long, saber-like wings remain motionless, and an occasional twist of the tail or turn of the head is the only visible movement as the three-kilogram bird exploits the force of surface wind currents, its solemn eyes turning gravely from side to side in keenest watchfulness.

An albatross is obviously at home in its native element, hitchhiking on the wind at sea. It can readily alight on water, braking to a stop in a controlled crash. Taking off is more difficult: the wings are spread, the neck is outstretched, and the webbed feet paddle furiously at full speed as the bird heads into the wind to become airborne. Laysan albatrosses that are returning to their breeding islands often fly in small groups just above the horizon in what appears to be battle formation. They fooled lookouts on Midway during World War II more than once, sending American troops into attack positions.

Albatrosses have long been part of the legends of Western blue-water sailors, who have watched them follow their ships since at least the days of Magellan. These curious birds no doubt followed ancient Hawaiians navigating their outrigger canoes between the main islands or on offshore fishing trips. Superstitious sailors thought that albatrosses were the spirits of seamen blown overboard during storms, and believed they portended wind and fog. Samuel Taylor Coleridge's early-nineteenth-century *Rime of the Ancient Mariner* coined the expression "an albatross around the neck" with its tale of a man who was first chastised for killing the bird that made the stormwinds blow, and later praised for destroying the bird that brought the fog.

The family name for all thirteen of the world's albatrosses, Diomedeidae, comes from Diomedes, the Greek hero of the Trojan War. The gods exiled Diomedes to an island in the Adriatic Sea and turned his companions into birds resembling swans. The word "albatross" is a corruption of the Portuguese and Spanish *alcatraz*, or pelican. Pelicans were the only large white birds that Portuguese navigators knew when they first encountered albatrosses off the southern African coast in the fifteenth century. Their bizarre courtship antics at breeding colonies have earned them the English and Japanese epithets "gooney birds" and *aho-dori* (fool birds). Anyone who has had more than a passing encounter with these magnificent creatures understands that such pejorative names are ill deserved.

Laysan, black-footed, and short-tailed (Steller's) albatrosses have wingspans that exceed two meters and are by far the largest seabirds in the North Pacific. North Pacific albatrosses are small in comparison with the wandering and royal albatrosses of the southern ocean, which have wingspans nearly twice as wide. Nine albatross species are confined to the Southern Hemisphere, where they are commonly observed in the Roaring Forties and Furious Fifties of the southern latitudes. Because a becalmed albatross is grounded, albatrosses in each hemisphere are essentially barred from the other by the equatorial doldrums. The thirteenth species, the waved albatross, is the only truly tropical species. Its breeding is restricted to La Plata Island, off Ecuador, and Española, a small volcanic cone in the Galápagos. Waved albatrosses "winter" in the cold-water upwellings of the Humboldt current.

The sizes and plumages of the three North Pacific albatrosses make field identification of adults easy, yet the plumages of some juvenile black-foots and short-tails are quite similar, and the occasional hybrid between a Laysan and a black-foot can create confusion. Black-foots are entirely sooty-brown except for small white areas at the base of the bill and below the eye. Older black-foots develop increasing amounts of white on their rumps. Laysans and short-tails have somewhat similar white plumages. Although George Steller discovered short-tails during his travels with Vitus Bering in Kamchatka and the Bering Sea in the 1740s, Laysans and short-tails were not recognized to be separate species until the nineteenth century. Laysans are one of the smallest albatrosses. They are mostly white but have sooty-brown tails, backs, and dorsal wing surfaces. Laysan wing linings are mostly white but are bordered with black and have two dark patches. Laysans have small black spots in front of the eye and gray cheeks. Short-tails have huge, distinctive pink bills and are the largest North Pacific albatrosses. In their definitive adult plumage, short-tails are mostly white with a yellow-buff wash on the head and the back of the neck. The tips of their wings and tails are dark brown. Laysans and black-foots have contrasting temperaments. Laysans are usually gentle, placid birds that can easily be handled at their nests while black-foots are easily annoyed when disturbed, snapping aggressively at human intruders and even engaging in projectile vomiting.

Albatrosses are long-lived. About half of the Laysans live to twelve years, and one individual has survived to the ripe age of forty-three. Interestingly, the survival rates of Laysans decline somewhat when they begin to breed and then increase after the fourteenth year. The increased mortality is attributed to the stress of nesting, especially for females. After the age of fourteen, surviving Laysans are established breeders and can better cope with nesting. About one-fifth of the experienced adults in a typical year take a rest and do not reproduce.

Distribution and Abundance

Today the breeding of Laysan and black-footed albatrosses is virtually confined to the Northwestern Hawaiian Islands, although Laysans have increased dramatically in the main islands since the mid-1980s and may number a thousand birds. Nesting has been attempted on Kauai, Oahu, Molokai, Niihau, and Moku Manu, but successes are still few. Since the mid-1970s a few pairs, including a bird that had been banded as a chick on Pearl and Hermes Reef, have begun to reestablish former colonies in the Bonins (Ogasawaras). In the mid-1980s a new colony of Laysans was discovered on Guadalupe Island, off central Baja California, and another colony may be forming on San Benedicto Island in the Revilla Gigedos, west of central Mexico.

In Japan, black-foots have small colonies in the Senkakus, in the Bonins, and on Torishima (Bird Island) in the Izu Islands. Torishima, some 500 kilometers south of Japan, is renowned for its importance to the endangered short-tailed albatrosses. For many years Torishima had the only known short-tail colony in the world. The eminent Japanese ornithologist Yoshimaro Yamashina estimated that feather hunters killed 5 million short-tails in the late nineteenth and early twentieth centuries, thereby eliminating them from the southern Ryukyus, Pescadores, Daitos, northern Bonins, and ten additional colonies in the southern Izus. Early naturalists in Alaska, such as William Dall and Otto von Kotzebue, observed large numbers of short-tails in the interisland passages of the Aleutians and believed that they bred there, but no nests have ever been found.

Torishima is an active volcano whose eruptions caused extensive damage in 1903 (the entire human population was killed), 1939 (the short-tails' old breeding grounds were buried under ten to thirty meters of lava), and 1941 (the mooring cove was buried by lava, so that landings became difficult). Torishima may erupt again at any time and biologists have hoped for a second, more stable colony. Fortunately, there is cause for cautious optimism. At least thirty-five short-tails visit Minami-Kojima, Senkaku Retto (Diaoyu Dao), in the southern Ryukyus, and in 1988 the Japanese naturalist Hiroshi Hasegawa confirmed the presence of a handful of chicks. Individual short-tails are also observed with increasing regularity in the Northwestern Hawaiians and may someday breed there. A short-tail was seen on Midway in 1938–39 and another in 1940. One persistent bird that was banded on Torishima as a fledgling in March 1964 spent mid-November to late March on Sand Island, Midway, virtually every year from 1972 to 1983 before disappearing. Another short-tail apparently took up winter residence on Midway in 1984–85 and has returned several seasons. Other short-tails have been observed at Tern Island, French Frigate Shoals, and Laysan. During the summer of 1981 at least three short-tails made landfalls on Midway and Tern.

The 2.5 million Laysans in Hawaii vastly outnumber the 200,000 black-foots. Most of the almost 400,000 Laysan pairs nest on Midway and Laysan. The world breeding population of black-foots is about 50,000 pairs, with the majority nesting on Laysan, Pearl and Hermes Reef, and Midway. Because albatrosses spend most of their early years at sea and because it is difficult to census them accurately on land, such population estimates must be taken with a grain of salt. Short-tails number about 400 individuals and are steadily increasing. The vast population of albatrosses has a remarkable effect on the marine ecosystem. The U.S. Fish and Wildlife Service estimates that Hawaiian albatrosses consume over a quarter million metric tons of marine resources each year, the equivalent of about one-tenth of the total tonnage of all commercial and recreational fisheries in the United States.

Albatrosses at Sea

Hawaiian albatrosses graze over large areas of cool water north of the archipelago. They probably remain within a few hundred kilometers of the islands during the breeding months, but during summer disperse throughout the cold waters of the northwestern and northeastern Pacific (Figure 15). Laysans and black-foots usually stay twenty or thirty kilometers offshore during the nonbreeding season, but short-tails often come within sight of land. Hawaiian albatrosses rely on their ability to fly great distances from their breeding islands to exploit more productive waters where nutrient levels are higher. There is some evidence that albatrosses have a feeding territory during nonbreeding months. Black-foots, which are attracted to ships by the galley waste and offal, follow a ship for only a few hours. This observation may imply a territory with a radius of about fifty kilometers. Laysans and black-foots sometimes perform parts of their courtship dances at sea, but such behavior is atypical.

Albatrosses spend most of their daylight hours on the wing rather than sitting on the water. They are attracted to any floating object, even plastic, and will settle onto the water in large assemblies when food is available. Black-foots are far more aggressive than Laysans and emit shrill screams as they quarrel over a ship's garbage. They are so notorious for mooching food that sailors call them

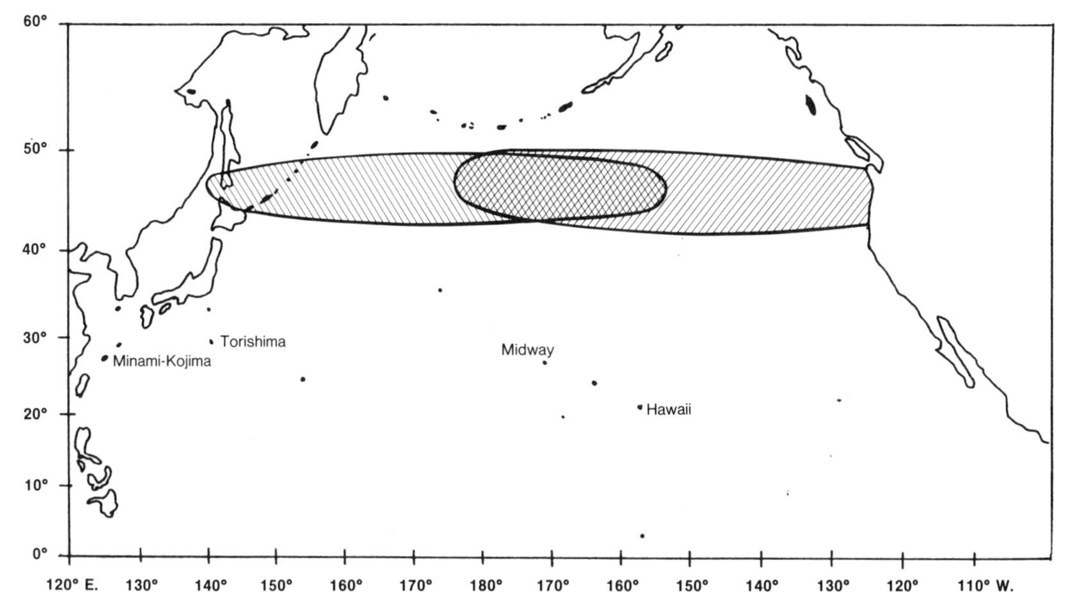

Figure 15. July ranges of adult Laysan albatrosses (northwest Pacific) and black-footed albatrosses (northeast Pacific)

feathered pigs. Albatrosses fly too fast to be able to forage on the wing, so all species must settle onto the ocean to feed, dipping their heads and bills into surface waters. Occasionally an albatross will submerge its whole body in an effort to snatch a particularly worthy item, but anything more than a body length below the surface is essentially out of these birds' grasp. Their long, powerful bills consist of several horny plates that are hooked at the tip (Figure 9*q*) to facilitate the seizing of slippery, wiggling prey.

Most albatrosses do much of their feeding during darkness or twilight. The retinas of Laysans possess high levels of rhodopsin, which aids nocturnal vision. Black-foots have far less. Such differences seem to account for some of the variations in diet between these species. While all albatrosses feed on squid, fish, and crustaceans, black-foots in Hawaii are unique in eating large quantities of flyingfish eggs. The eggs float on the surface of the ocean, often attached to flotsam, and are readily located by black-foots during daylight hours. Laysans feed primarily on squid that possess light organs and that migrate to the surface from deep water during darkness. Laysans are better adapted than black-foots to find such prey. Hawaiian albatrosses apparently have evolved to feed at different times, so that competition between them for food is reduced.

One puzzling feature of the diets of Hawaiian albatrosses is the frequent occurrence of certain deep-water crustaceans. One amphipod (*Eurythenes gryllus*) that occurs at abyssal depths up to 6,500 meters was found in a Laysan albatross's stomach. A 340-millimeter specimen of another amphipod (*Alicella gigantea*) that lives at similar depths—the largest amphipod known to science—was found in a black-foot. Some isopods and mysids eaten by albatrosses generally live in waters thousands of meters deep. How do surface-feeding albatrosses encounter such creatures? One favored hypothesis is that such crustaceans, like squid, migrate to the surface at night. Other biologists suggest that such organisms drift to the surface attached to the dead prey on which they feed. Still others believe that sperm whale feces or vomit makes such denizens of the deep available in surface waters where birds may eat them. Sperm whales periodically empty their stomachs of squid beaks, which clog their guts, and are known to regurgitate their stomach contents when they are chased by whalers. Marine biologists observe freshly vomited squids floating on the sea surface with some regularity, and many types of seabirds are attracted to them. Another interesting feature of albatrosses' food habits in Hawaii is their frequent ingestion of flotsam. Plastic, charcoal, fishing line, and Kellogg's Rice Krispies wrappers have been found in albatrosses' stomachs. Naturally occurring items also turn up, including squid beaks, pumice, and an occasional kukui nut.

The marine environment can be fickle, and albatrosses have occasional bad years for reasons that biologists do not fully understand. Studies by Harvey Fisher on Midway indicate that 1964–65 and 1968–69 were very poor years for Laysans, and only about half the normal number of birds nested. Severe storms in December 1964 contributed to a massive early desertion of eggs. Interestingly, storms associated with El Niño caused similar problems in the Galápagos just six months later. Adult mortality was not abnormal at either Midway or the Galápagos, and the birds returned in succeeding years. Apparently the conditions for foraging were poor and many adults sensibly choose to rest that year. Severe die-offs of fledglings on Pearl and Hermes Reef in 1933, 1963, and 1978 may have been related to the availability of food. Biologists rarely visit Pearl and Hermes, and such extraordinary mortality may occur much more often than we know. The boom and bust years of albatrosses may indicate the general productivity of the marine environment.

Albatrosses on Land

Except for sooty albatrosses in the Southern Ocean, all albatrosses breed in large colonies. Although the true home of an albatross is the sea, it must return to terra firma to breed. An albatross arriving home after months at sea often skids to an awkward, rolling sprawl on land. Its large webbed feet are more of a hindrance than an asset ashore, and birds seem clumsy when they waddle and plod from landing areas to nesting territories. Some biologists suggest that because Hawaiian albatrosses breed at the same time of year as their Southern Hemisphere counterparts, they are somehow bound to the breeding season of albatrosses five thousand kilometers to the south and millions of years distant in evolution. Antarctic and subantarctic albatrosses breed between September and November to minimize the effects of severe weather and to coordinate chick-rearing with the maximum period of food availability near their colonies. North Pacific albatrosses have evolved to nest in late fall because of weather and food conditions in the Northwestern Hawaiian Islands. The fact that Hawaiian and Antarctic albatrosses breed at vaguely similar times of the year is pure coincidence.

Arrival at the Colony

Hawaiian albatrosses have been extensively studied on the Midway Islands. Like the coming of the swallows to Capistrano, their arrival ashore is predictable and precipitous, and the quiet islands are transformed into clacking, groaning, mooing barnyards of goose-sized seabirds. The first black-foots arrive in mid-October, and the vanguard of the Laysans follows about two weeks later, in early November. Laysans in the main islands have similar schedules. After a month of rapid growth, the population of each species at the colony stabilizes. All North Pacific albatrosses reproduce synchronously: laying, feeding, and fledging occur at the same time each year.

The homing mechanism that enables an albatross to find its own island in the vast North Pacific is still a puzzle. Rarely will a Laysan albatross set foot on an island where it did not hatch. These birds manage to return not only to their home island but to the identical patch of sand or grass where they hatched. This tenacity makes the founding of a new colony or the repopulation of an extirpated one very rare. Marcus (Minami Torishima) and Wake still lack breeding albatrosses almost a half century after colonies there were destroyed, although the pioneering birds on the main islands are ample evidence that some colonies can be revived. Homing experiments were conducted in which albatrosses were flown by military transport planes from Midway to Oahu, Japan, Kwajalein Atoll, the Philippines, and Whidby Island, Washington. Fourteen of eighteen birds found their way back to Midway, and one bird made its 5,000-kilometer return trip in just ten days.

Male Laysans arrive at the colony before females. While waiting for the females to arrive, they usually sleep with their heads on their backs and beaks stuck into their scapulars. Occasionally they thrust their beaks into the air and give melancholy cries, which biologists have aptly named sky calls. Birds arrive by age group, the older, more experienced birds arriving before the younger. Arrival by age group enhances the chance that pairs will form between birds of similar ages. This is an important strategy for the population because it reduces the possibility of mismatching and untimely disruptions of pairs when older birds die.

Each species prefers its own nesting habitat. Black-foots tend to nest in open, wind-swept, sandy areas near shore, whereas Laysans choose vegetated, inland areas. Both species can breed on rocky strata such as Kaula, Nihoa, and Gardner Pinnacles. A first-time visitor to Midway may initially be confused because some black-foot colonies are inland. Extensive land modifications after World War II reshaped the beaches when portions of the reef were dredged and filled to expand the island. Black-foots tenaciously return to long-held breeding areas, even though they are now vegetated and hundreds of meters from shore. After more than forty years, very few birds have pioneered nest sites in the landfill areas.

Courtship and Incubation

Albatrosses court potential mates by stereotyped dances, bows, bill snaps, and nasal groans. The purpose of such complex and bizarre activities is for juvenile birds to establish, not maintain, pair bonds. The dances of Laysans and black-foots are similar, but black-foots have louder, harsher voices and their tempos are faster. Most courtship takes place between mid-March and mid-May, when birds that will breed during subsequent years attempt to establish territories and find mates. During this period of intense courtship displays, most breeding adults are at sea searching out and acquiring food for their growing young. The vast majority of Hawaiian albatrosses begin breeding between seven and nine years of age.

Albatrosses have a more sustained and intricate behavioral dialogue than almost any other bird, with each action eliciting a predictable response from the partner. Walter K. Fisher vividly described the courtship displays of Laysan albatrosses during his expedition to Laysan in 1903:

> Two albatrosses approach each other bowing profoundly and stepping rather heavily. They circle around each other nodding solemnly all the time. Next they fence a little, crossing bills and whetting them together, pecking meanwhile, and dropping stiff little bows. Suddenly one lifts its closed wing and nibbles at the feathers underneath, or, rarely, if in a hurry, merely turns its head and tucks its bill under its wing. The other bird during this short performance assumes a statuesque pose and either looks mechanically from side to side or snaps its bill loudly a few times. Then the first bird bows once and, pointing its head and beak straight upward, rises on its toes, puffs out its breast, and utters a prolonged nasal groan, the other bird snapping its bill loudly and rapidly at the same time.
>
> Sometimes both birds raise their heads in the air and either one or both utter the indescribable and ridiculous bovine groan. When they have finished, they begin bowing at each other again, almost always rapidly and alternatively, and presently repeat the performance, the birds reversing their role in the game, or not. There is no hard and fast order to these antics, which the seamen of the *Albatross* rather aptly called a "cake walk," but many variations occur. The majority of cases, however, follow the sequence I have indicated. Sometimes three engage in the play, one dividing its attention between two. They are always most polite, never losing their temper or offering any violence. The whole affair partakes of the nature of a snappy drill, and is more or less mechanical.

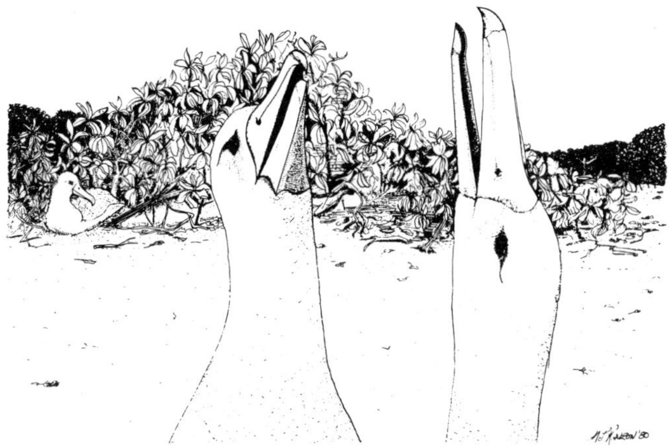

Laysan albatrosses sky-pointing

Adult albatrosses arrive at the colony with heavy layers of fat acquired during the summer in the North Pacific. Before the female has made landfall she has already begun to deposit her eggyolk, stimulated by environmental factors. Albatrosses rarely "divorce," and unless her mate has died, she reunites with him on his small territory. Birds paired during previous years engage in few of the elaborate displays that are characteristic of courting birds. After many months of separation at sea their greetings are confined to bill touching and nibbling. An experienced pair will often mate within a few hours of reunion, always within the male's territory.

Both birds return to sea to feed for several weeks while the egg continues to develop. About twenty-four hours before the female lays her egg, she returns to select a nest site within the territory. She does minor work in scratching out a hollow, but the male does most of the actual construction during his first incubation span. The rim of a Laysan nest may be as high as an adult's belly and serves to protect nestlings from drowning in the pools of rainwater that accumulate during winter storms. Black-foot nests are usually lower and shallower, often consisting of a mere depression in the sand. Male and female albatrosses actually spend very little of their lives together. They are entirely apart at sea during the summer months, and the chores of rearing a chick allow them to spend perhaps five or ten days together each breeding season.

Most Laysans lay a single 280-gram egg between late November and mid-December, and the zenith of black-foot laying occurs a week to ten days earlier. Eggs are white with a chalky appearance. Older, more experienced birds lay first and have the greatest chance of raising young successfully. Both males and females incubate, using a featherless incubation patch to control temperature and humidity for the growing embryo. Most pairs have five or six incubation spans, beginning with a two-day span by the female. After the male returns from the sea to relieve her, she departs the island to fatten up. The male's first incubation shift often lasts as long as three weeks, and consequently visitors to a colony in early December will find that incubating male Laysans outnumber females 15 to 1. As incubation proceeds, the bouts shorten to about one week.

Incubating birds slow their bodily functions and face into the wind to prevent ruffling of feathers and hypothermia. They do occasional housekeeping to remove sand that has been blown into the nest depression, but nest maintenance ceases when the egg hatches. Albatrosses carefully jet their cloacal effluvia beyond the rim of the nest, and the radiating white spokes fertilize the lawns of Midway. Parents have a nest relief ceremony when one bird returns to relieve its mate, which serves to strengthen the pair bond. It can take considerable coaxing on the part of a mate to persuade the incubating bird to leave the egg. Throughout incubation, parents talk to the egg with a three syllable *eh-eh-eh*, which probably helps the developing chick to recognize the voices of its parents. The sixty-five-day incubation period serves as a forced diet—males lose as much as a quarter of their weight, most of which they regain within a few weeks back at sea.

Raising the Chick

Black-foot chicks begin to hatch in mid-January, just before the Laysans. Hatching success varies widely from year to year, but tends to run between 60 and 80 percent. A wet, gray hatchling emerges from the shell several days after the appearance of a star fracture and adults alternate at brooding the downy young for three or four weeks. In their early stages, young are fed at least once each day. At this time the parents must depend on local food resources because they cannot commute long distances from the colony. For the first few days, hatchlings are fed exclusively a putrid-smelling stomach oil, which diminishes in importance as they grow older and receive more solid food. A biologist working with young albatrosses cannot avoid getting some of the oil on his clothes, and no detergent yet invented will completely remove its pungent odor. Stomach oil is common with all tubenoses. Its origin has long been debated, but biochemists now agree that most of the oil's fatty acids, alcohols, and wax esters are derived from food rather than from the birds' glandular secretions. The lipids in the oil are strikingly similar to those found in deep-water crustaceans, fishes, and squid. Stomach oil is essentially a concentration several times over of the energy found in prey and is an efficient means of transporting food from distant feeding grounds to the colony. Oil, when metabolized, is also a source of water.

Adults feed their young only in the immediate vicinity of the nest site. A parent returning from the sea is immediately besieged by its hungry chick, which vehemently nibbles at the side of its bill, begging pathetically until the parent is stimulated to disgorge its meal. When the adult opens its beak, the young inserts its own bill crosswise, bolting down the seafood soup with relish. A young albatross can be virtually immobilized after devouring a large meal. During March and April, the Northwestern Hawaiian Islands reek with rotting food items that are too large for young birds to ingest and the uric acid stench that is characteristic of all seabird colonies.

A fairly high proportion of albatross chicks that hatch survive to fledge, as many as nine birds in ten during a good year. Storms, heat, and starvation are the primary causes of death among chicks, although avian pox and droopwing take a few. Severe winter sandstorms can bury young black-foots alive in their nests on the beach, and they instinctively protect themselves by continually kicking out encroaching sand. High surf caused the loss of nearly half of the black-foot chicks at Laysan in 1958, more than four hundred Laysan nests on Midway in 1962, and hundreds of black-foot nests at French Frigate Shoals in 1982. At Laysan, heavy winter rainfall in 1913 caused the lagoon to rise and drowned some 3,000 nests in the interior. Heat stress poses a problem for Hawaiian birds which Southern Hemisphere albatrosses rarely encounter. Chicks and even adults cool themselves by sitting back and balancing on their heels with their webbed feet raised and spread, avoiding contact with the hot ground and permitting heat loss from their feet. Chicks place their backs to the sun and their feet in their own shadows to obtain the full benefit of onshore breezes. Nestlings will wander short distances from their nest sites to seek shade. Most starvation occurs when one parent dies during the nesting cycle and the mate is unable on its own to provide sufficient food. Not only does the death of a parent result in the loss of young during that season, but the surviving adult may lose the subsequent breeding season when it attempts to form a new pair bond.

By mid-May, many Laysan chicks weigh 3.2 kilograms, a quarter more than their fathers. As summer approaches, parents become less attentive and visit the colony infrequently. The young gradually begin to lose weight, living largely off fat reserves as they replace down with feathers. Once they have attained adult proportions, fledglings take advantage of every rise in the wind to exercise, flapping their wings awkwardly in the air. A visitor to a colony in late May or June is treated to the marvelous sight of fields of fledglings spreading and waving their wobbly wings. One can almost see the disbelief in their eyes when they jump off the ground and find themselves airborne on their first short flight. Parents continue to shuttle in food during the period of test flights, but the frequency drops to every third day or so. During June, albatross colonies are littered with pellets of squid beaks and opaque eye lenses, apparently disgorged by young birds lightening the load before going to sea.

Black-foots grow much faster than Laysans, taking only about 140 days rather than 165 from hatching to fledging. Apparently black-foot diets are more nutritious, especially the flyingfish eggs, which contain three times the concentration of calories in squid. Successful Laysan fledglings begin to depart in mid-June and most are gone by the end of July. They usually weigh about two kilograms, a third less than their maximum weight in May. Fledglings often walk to the shore with groups of neighbors, and most survivors depart within two days of reaching the shore.

Venturing to sea is dangerous. Many young have drowned at Laysan attempting to go to sea during heavy inshore surf. Tiger sharks patrol the inshore waters of all breeding islands in late June and July and eat one bird in ten, many of which seem to be healthy, active fledglings. Other young birds die of exhaustion, exposure, or starvation just offshore and may also become shark prey. Those that reach open water are alone against the sea—they must learn to obtain food without the assistance of their parents.

Conservation

Albatrosses are resilient and have endured at disturbed colonies despite numerous problems created by humans. On Midway, they pay little attention to cars and trucks that rumble close to their nests. The ability of Laysans to maintain nesting on lawns and golf courses on Midway and to ignore humans has been an advantage. A more timid species might long ago have abandoned nesting, yet Laysans and black-foots retain one of their largest colonies on a small atoll that housed as many as 15,000 people during World War II and more than 3,000 until the 1970s.

It is the vogue to point to the problems that humans create for wildlife. It is true that buildings, lights, antenna wires, and even introduced ironwood trees have created barriers that kill many albatrosses on Midway each year. Daylight landings and takeoffs of airplanes on Midway began with Pan American Airways' flying boat service in 1935 and continue today with C-141 jet aircraft. Thousands of albatrosses have been killed in collisions with airplanes and tens of thousands were removed in an attempt to reduce the number of such collisions. Yet the introduction of soil and grass to Midway has added much nesting habitat. Albatross populations on Midway today are much larger than populations reported by the earliest visitors. In 1891, Sand Island well deserved its name, and few albatrosses nested in the dunes. Shipwrecked sailors may have eaten many albatrosses before anyone recorded estimates of bird populations, but it seems unlikely that humans could have consumed as many as 400,000 albatrosses, the most recent estimate of breeding birds there. Today there are probably more albatrosses on Midway than there ever were before, and human interaction over time has been a net benefit to albatrosses there.

All three North Pacific albatrosses suffered seriously from the millinery trade during the late nineteenth and early twentieth centuries, when short-tails almost became extinct. Albatross feathers were more valuable than those of other seabirds, and hunters raided colonies at Midway, Laysan, and Lisianski. Other colonies in the North Pacific, such as the Volcano (Iwo) Islands, Wake, and Marcus, have never recovered and still lacked a single breeding pair in the mid-1980s. Marcus once had an estimated population of one million Laysan albatrosses. Hawaiian colonies have generally attained their former levels, except that populations on Lisianski have apparently remained depressed. Rabbits introduced to Lisianski early in the twentieth century may have altered the vegetation and adversely affected nesting habitat.

Since the late 1970s a minor miracle has been occurring—Laysan albatrosses have begun to recolonize the main islands. Although bones in subfossil deposits indicate that albatrosses have probably inhabited the main islands during the human era, until recently historical records were limited to the colony of several hundred pairs on Niihau. George C. Munro believed that a pair of Laysans that came ashore at Koloa, Kauai, in 1945 were the first on any of the main islands. A black-foot found soon thereafter on Kailua beach was also unique. Until the 1950s, Laysans were rarely seen in the waters offshore the main islands, although a few pair began to breed on Moku Manu in 1947.

Courting Laysans began to turn up on Kauai in 1975 and black-foots occasionally landed, especially on Moku 'Ae'ae, offshore Kilauea Point. In February 1976 the first Laysan nestling was discovered on Crater Hill; it died in midsummer. By 1979, Laysans were alighting at Waikane and Kahuku on Oahu and nests were common at Barking Sands, Kauai. By the late 1980s, several successful Laysan chicks were being raised each year at Kilauea Point and nesting was attempted at nine other locations on Kauai. Groups of thirty or more Laysans attempted to nest along Oahu's north shore from Kaneohe Marine Corps Air Station to Ka'ena Point. Displaying birds have been observed from near sea level to an elevation of 500 meters at Ka'ena Point. An apparent nest was found on a coastal bluff overlooking Kawakiu Niu Bay, Molokai. Scores of birds have been killed by wild dogs at Dillingham Air Field, Oahu, and Barking Sands, Kauai, and feral pigs are apparently exacting a serious toll at the Niihau colony. If Laysans continue their attempts to colonize the main islands, controls may become necessary near airports to ensure airline safety.

Albatrosses can be harmed by the introduction of alien creatures on their breeding islands. Polynesian rats occasionally attack incubating albatrosses on Kure, leaving large, gaping wounds in their backs. Mosquitos introduced on Midway have been implicated in the transmission of avian pox, a viral disease that causes extensive, gruesome facial lesions on young albatrosses. Well-meaning but overzealous residents of Midway sometimes remove diseased birds from the colony unnecessarily in an attempt to slow the spread of the disease. Most birds actually survive this chickhood disease if they are left alone, but none survives removal from a nest site. Avian pox sometimes infects albatross nestlings on Kauai. Some albatrosses perish at sea when they are accidentially hooked by long-line tuna fishermen, and others drown during gillnet fishery operations that stripmine the North Pacific. Fisheries are not yet a major problem to albatross populations, but biologists must closely monitor developments in fishery technology to ensure that such problems remain minor.

Albatrosses are the largest birds in the North Pacific. Their intricate breeding behavior has long fascinated biologists, and they are an important component of the marine ecosystem, especially near their colonies. Short-tails are making an encouraging recovery after erroneously having been declared extinct in 1949. As indicators of the ever-changing conditions in the ocean, albatross stomach contents provide an index of the amount of nondegradable plastics that have been dumped into the sea. With wise management, their survival for future generations should be assured.

10 SHEARWATERS AND GADFLY PETRELS
Family Procellariidae

The weird, gurgling cries in the inky North Pacific night awakened me from a deep sleep. Groggily I realized that I was lying in a sleeping bag with some unfriendly rocks poking through the Ensolite pad into my rib cage and thighs. What creature could be making those horrible caterwauling sounds? Surrounding me and silhouetted against the Milky Way were marae, huge stone tablets from an ancient era. My colleagues were sprawled nearby in the clearing of what was once a Polynesian heiau on Annexation Hill, exhausted from a day of hauling gear ashore on Necker and then clambering up and down crumbling rock ledges and terraces in the late-May sun to survey the seabird colonies on this forlorn volcanic stack. I am not particularly superstitious, but before the pink sky faded into darkness I had taken the precaution of pouring a shot from my bottle of Jack Daniels onto the dust of our campsite. Long before I had seen a shaman perform a similar ceremony in upcountry Ghana, and a few words of apology to any lingering spirits of the ancient Polynesians seemed a reasonable precaution before my team of biologists made camp within the religious shrine. The screaming, wailing noises continued, apparently emanating from some rocks just beyond my reach. If anything could wake Necker's dead from their eternal sleep, this cacophony would.

My consciousness focusing, I determined which culprit must be making the catlike squalls. Rising from my sleeping bag, I reached into the rocks and pulled out a wiggling wedge-tailed shearwater. Like all wedge-tails, he was both strong for his size and ornery. He seized and twisted one of my fingers with his bill while scratching my forearm repeatedly with furious kicks. Walking to the edge of a nearby crest, I noticed that the *Townsend Cromwell* was safely anchored below in West Cove, its anchor lights a friendly beacon in the darkness. Without apology, I cast the shearwater off the cliff into the night. We were here to study and conserve seabirds, but I draw the line at a sleepless night. Returning to my makeshift bed, I scratched the welts on my arm where seabird ticks had earlier sucked an evening meal. Within half an hour, Pavarotti had returned and renewed his attempts to attract a mate with his harsh, bloodcurdling moans. By then it was a few minutes past midnight, and I realized that my thirty-first birthday had arrived. I repeated my attempts at banishment, but like a boomerang the bird returned again and again. The rising sun over the heiau a few hours later was an especially welcome sight.

All shearwaters and petrels are known for strange vocalizations at their colonies at night, but wedge-tailed and Newell's shearwaters are probably the loudest of the six species that nest in Hawaii. Shearwaters and petrels are highly pelagic, spending much more of their lives wandering the trackless oceans than caterwauling to mates on Hawaiian islands. Shearwaters are so named because sailors thought they sheared the water with their flitting, batlike flight when they swooped along the troughs of giant waves on virtually motionless wings. "Petrel" is probably a diminutive of Peter, an allusion to St. Peter's attempt to walk on water. The family name, Procellariidae, is derived from the Latin *procella* (storm) and refers to the association of shearwaters and petrels with storms at sea.

Depending on which taxonomy one uses, there are about fifty-five species and twelve genera in the family. The family is extremely successful, with members in all unfrozen salt waters of the earth. The genera *Pterodroma* (wing runner) and *Bulweria* (named after James Bulwer, a fellow of the Linnaean Society) are gadfly petrels. The genus *Puffinus* (named from an erroneous belief that shearwaters are related to puffins) includes most shearwaters. Procellariidae probably evolved primarily in the Southern Hemisphere, where three-quarters of the species breed and spend most of their lives. The earliest procellariids known from fossils are shearwaters of the Middle Oligocene, about 40 million years ago. Gadfly petrels, in contrast, are known from only the past 2 million years. Though the fossil record of petrels may be deficient because of their tendency to nest on quickly eroding islands that are poor substrates for fossils, it seems certain that they emerged after shearwaters.

Shearwaters and petrels are medium-sized birds with long, thin, tapering wings about a meter in length. Their bodies are more slender than those of albatrosses. All shearwaters and petrels have nostrils that open through horny tubes lying on top of the upper mandible. Their rather long, deeply grooved bills are hooked to facilitate the retention of prey and vary only in the proportion of length to depth. As a family, procellariids are drab. Most plumages are black, dark brown, or dark gray, and many tend to be gray above and lighter below. The sexes cannot be distinguished by size or plumage. Shearwater and petrel feathers usually have a musky odor that clings to burrows and museum skins long after the birds have departed or died.

Shearwaters are highly adapted for aquatic life but have sacrificed some diving efficiency for improved flight. Swimming is enhanced by laterally compressed legs, strong flattened tarsi, narrow pelvises, and the placement of legs far back on the body. Such adaptations promote efficient swimming and the underwater pursuit of prey powered by wings and webbed feet. Shearwaters have a higher specific gravity than gadfly petrels and possess dense, waterproofed plumage. Their wings are shorter than those of other procellariids but are still long and narrow enough to provide efficient, high-speed gliding flight. Tails are comparatively short and stiff. On land, shearwaters have some difficulty taking flight in the absence of wind.

Gadfly petrels, in contrast, do not swim or dive after prey. They swoop and soar, now and again beating their wings. They are usually smaller than shearwaters and their wings appear to be bent at the wrist. Their wing-loadings are lower than those of shearwaters, an aid in manuverability and prolonged gliding. Petrels detect prey on the wing and capture it close to the sea surface, by either dipping or settling onto the water to seize it (Figure 11). Their underwing pattern apparently disrupts the wing silhouette and may aid the capture of quick-reacting prey such as squid.

Although the difficulties of identifying shearwaters and gadfly petrels at sea are notorious, they are minimized in Hawaii, where only three shearwater and three gadfly petrel species breed. Wedge-tails, the largest of the Hawaiian shearwaters, have long, distinctly wedge-shaped tails that are distinguishable when they turn in flight. They have two distinct color phases. Light-phase birds are a dirty white on the underparts except for dark underwing margins and undertail coverts. The upperparts of fresh-plumaged birds are a brownish gray, which becomes browner with wear. Dark-phase birds are entirely sooty brown, and are seen primarily in the Southern Hemisphere. Bills are slaty gray and legs and feet are flesh-colored. Newell's shearwaters, considered by some taxonomists to be a subspecies of Townsend's shearwaters, have glossy black upperparts. The dark contrasts sharply with the pure white below, which extends well up the sides of the neck and flanks. Bills, legs, and feet are dark. Newell's shearwaters weigh 370 grams, slightly less than wedge-tails and slightly more than Christmas. Christmas shearwaters are almost uniformly sooty brown to black with dark-brown eyes, black bills, and dark feet. They are easily distinguished from wedge-tails by their smaller size and short, rounded tails.

Dark-rumped petrels, sometimes call Hawaiian petrels, weigh an average 434 grams and are the largest species in this family that breeds in Hawaii. Their upperparts, including nape, wings, back, rump, and wedge-shaped tail, are grayish black. Their white foreheads and cheeks can give them a white-headed appearance from a distance. Underparts are also white except for prominent dark margins on the underwings. Legs and feet are pink and bills are black. The plumages of Bonin petrels vary widely, depending on condition. In fresh plumage, Bonins have handsome silver-gray backs and sooty-gray heads, necks, and flight feathers. Foreheads and underparts are white. When plumage becomes worn, the light gray of the neck and back of Bonins darkens and merges with the gray of the head and neck. The forehead gradually changes from white to gray, and by summer the gray of the head extends down in a V toward the ebony bill. Bonin legs and feet are mostly flesh-colored. Bulwer's petrels are smaller than other gadfly petrels, almost as small as storm-petrels. They are uniformly sooty brown except for buff-colored bars that run diagonally across the upper wing surface. Legs, feet, and bills are a uniform black. In the field, Bulwer's petrels can be distinguished from similar sooty storm-petrels by their stouter bills and long, wedge-shaped tails.

Like most seabirds, petrels and shearwaters are long-lived. We can only guess the length of a normal life span because the birds destroy the aluminum and Monel bands that are used to determine their ages. No band subjected to annual burrow excavation can survive as many years as a shearwater or petrel. Bulwer's live at least twenty-two years, while Bonins, wedge-tails, and Christmas shearwaters have been proven to live only ten or eleven years. No doubt an intensive banding and rebanding program would yield records of much longer survival.

Distribution and Abundance

Wedge-tails are the most abundant tropical shearwaters in the North Pacific and breed widely in tropical and subtropical waters of the Indian and Pacific oceans. The western and eastern limits of their breeding distribution are Madagascar and the Revilla Gigedos, respectively. They breed on all Pacific island groups except Easter, Palau, the Tuvalus (Ellice Islands), the Gilberts, and the Cooks. Christmas shearwaters breed in relatively low numbers on many islands of the central Pacific between 30 degrees north latitude and 30 degrees south. They also breed on Henderson, Ducie, and Oeno, and in the Marquesa, Tuamotu, Austral, Line, and Phoenix groups. Christmas shearwaters have been extirpated from the western Pacific colonies on the Bonins (Ogasawaras), Marcus (Minami Torishima), and Wake. Newell's shearwaters are endemic to Hawaii.

Dark-rumped petrels breed only in Hawaii and the Galápagos, each population constituting a separate endangered subspecies. The breeding range of Bonins is restricted to the Northwestern Hawaiians, Bonins, and Volcanos (Iwos). Bulwer's petrels breed in the subtropical Pacific and Atlantic, including Johnston Atoll, the Hawaiians, Marquesas, Phoenix, Bonins, Volcanos, Izus, Ryukyus, Azores, Salvages, Canaries, Cape Verdes, and Desertas (offshore Madeira).

Wedge-tailed and Christmas shearwaters are common on Kaula and most of the Northwestern Hawaiians (Tables 1 and 2). In the main islands, wedge-tails breed on Kauai, Lehua, Maui (Waihe'e Point), and the islets and seastacks offshore Lanai (Pu'u Pehe, Po'opo'o, Ki'ei), Maui (Molokini, Hulu, Moke'ehia), Molokai (Mokuho'oniki), and Oahu (all except Mokolea, where burrowing is impossible). Dogs and cats probably kill the pioneering wedge-tails that sometimes attempt to nest at Black Point, Oahu. In the main islands, Christmas shearwaters are known only at Moku Manu and Lehua. Newell's shearwaters breed on the forested slopes of Kauai between 150 and 800 meters (Figure 8). Observations of fledglings and calling adults indicate that small colonies still exist on other main islands. Bulwer's petrels generally breed on the same islands as wedge-tails except those inhabited by rats (Midway, Kure, Mokoli'i, Moku'auia). Bulwer's called from the low cliffs at the southeast end of Oahu in the 1940s. Bonins breed on French Frigate Shoals and all islands north except Gardner Pinnacles. The only remaining nesting areas for dark-rumped petrels which have been located are the volcanic slopes of Haleakala Crater (Figure 4), but fledglings have been recovered on Kauai, Lanai, and Hawaii, and small colonies certainly exist there.

Most shearwaters and petrels spend many years at sea before breeding and thus at any one time much of any population is observed only at sea. They do not make their first landfall until they reach about five years of age and first breed several years later. An estimated half of the 400,000 Bulwer's and 1.3 million Bonins that visit the Hawaiian breeding colonies are nonbreeders. Over half of the 14,000 Christmas shearwaters and two-thirds of the 1.8 million wedge-tails are believed to be nonbreeders. The breeding population of dark-rumps has declined to between 400 and 600 pairs, but the Newell's population is much greater, numbering between 4,000 and 6,000 pairs. The U.S. Fish and Wildlife Service and the State of Hawaii have listed them as endangered and threatened species, respectively. The structures of these populations have been so distorted by human disturbance that nonbreeding populations cannot be estimated.

Shearwaters and Gadfly Petrels at Sea

Shearwaters and petrels spend by far the greater part of their lives at sea, often undergoing extensive migrations. Short-tailed shearwaters breed near Bass Strait, Australia, and cross Hawaiian waters during annual transpacific migrations to Kotzebue Sound, just north of the Arctic Circle in Alaska. Hawaiian wedge-tails and Bonins apparently migrate to the Gulf of Panama and Japan, respectively, after their young have fledged. During breeding season, most shearwaters and gadfly petrels feed within eighty kilometers or so of their colonies in the Hawaiian Islands. Newell's shearwaters may have a major feeding ground in the waters bounded by Kauai, Niihau, and Kaula.

At sea, Hawaiian shearwaters are most active during the day. Christmas and wedge-tails feed primarily in association with skipjack tuna and other predatory fish that usually feed during daylight hours. Less is known about the habits of Newell's, but they seem to forage similarly. Christmas and wedge-tails consume mostly larval forms of goatfish, mackerel scad, flyingfish, squirrelfish, and flying squid which are driven to the surface by fish schools. Christmas shearwaters, like black-footed albatrosses, have low levels of rhodopsin in their eyes and probably see poorly in the dark. Wedge-tails and Newell's probably also have poor nocturnal vision, but wedge-tails sometimes feed at night when the moon is full.

Gadfly petrels are fundamentally nocturnal both at sea and on land. Bonins, Bulwer's, and dark-rumps forage largely at night or twilight. The retinas of Bonins, like those of Laysan albatrosses, have high levels of rhodopsin, which aids nocturnal vision. All gadfly petrels that nest in Hawaii eat large amounts of juvenile flying squid. They also consume fishes that inhabit deep waters by day and surface at night: lanternfish, hatchetfish, bristlemouths. On bright moonlit nights petrels are less active at colonies and spend more time at sea. This phenomenon is sometimes ascribed to an innate fear of terrestrial predators, but few gadfly petrels have evolved on islands where predators are present. Michael Imber of the New Zealand Wildlife Service has suggested that gadfly petrels spend more time than usual searching for prey on moonlit nights because fishing conditions are poor. Under a full moon, vertically migrating fish and squid remain deeper in the water column and are less available to birds.

Like albatrosses, gadfly petrels consume some puzzling organisms. Deep-dwelling crustaceans such as Oxycephalidae (an amphipod) and *Anuropus* (an isopod) occasionally turn up in Bonin and dark-rump stomachs. Gadfly petrels feed in the surface waters of the water column, so how can they obtain such organisms? The answer may lie in their ability to scavenge—petrels are well known for feeding on naturally occurring waste such as dead whales and squid, and no doubt will feed on any dead organism that floats to the surface from midwater. They avoid humans' garbage, however.

Gadfly petrels and shearwaters use their sense of smell to locate food. Scientists have used animal fats to lure procellariiforms to research vessels since at least the turn of the twentieth century. Though birds as a group are regarded as oblivious of odors or insensitive to them, seabirds have prominent olfactory bulbs in their brains similar to those of mammals. The characteristic odors associated with petrel and shearwater burrows may be related to their sense of smell. Wedge-tails deposit excrement just outside their burrow entrances, which may serve as an odorous label. Some colonies are so permeated with tubenose scent that sailors many kilometers downwind can smell it.

Gadfly petrels and shearwaters consume over 50 million tons of marine organisms each year in Hawaii, which account for about one-eighth of all the prey consumed by the Hawaiian seabird community. Most of this consumption is attributed to Bonins and wedge-tails; it is fair to state that the collective impact of Bulwer's and dark-rumped petrels and Newell's and Christmas shearwaters on the marine ecosystem in Hawaii is trivial.

Shearwaters and Gadfly Petrels on Land

The true home of shearwaters and gadfly petrels, like that of albatrosses, is the ocean. The very adaptations that have shaped their marine existence have undermined their ability to function ashore. Shearwaters are particularly awkward on land, with legs placed so far to the rear that they cannot walk. When ashore, wedge-tailed and Christmas shearwaters prefer to sit, and undertake short waddling runs only if they are disturbed. Bonin petrels walk with a rolling gait. They hold their bodies low and horizontal, with head and neck outstretched and wings folded to their bodies. Newell's have a particularly difficult time landing. Their colonies are now mostly restricted to forested areas in the windward mountains of Kauai, under which are dense, impenetrable stands of uluhe ferns. Birds returning to their burrows crash into the treetops and tumble from branch to branch into the dense understory of the fern. They can use their claws and wings to lift their heads clear, but occasionally individuals strangle in tree crotches. Newell's hollow out tunnels under the ferns, through which they crawl to their burrows, sometimes a thirty-minute trip.

Shearwaters and gadfly petrels developed the basic features of their reproductive cycles early in their evolution and have retained them with little modification under the varied ecological conditions in which the species now live, despite the fact that environmental pressures have created a wide range of body sizes and feeding patterns. In many important ways, their life histories on land parallel those of the albatrosses: they first breed relatively late in life and have long incubation bouts.

Breeding Season and Nest Sites

Hawaiian shearwaters and gadfly petrels breed synchronously and arrive early in the nesting season to build or reclaim their nest sites. With the exception of Bonins, all Hawaiian species return to their nesting colonies between February and April and depart in October and November. Bonins return in August and September, do not lay until January, and leave their colonies in June. Most dark-rumps are now restricted to nest sites along the heavily eroded west rim of Haleakala Crater, Maui, where temperatures are cool. The locations of their colonies, between two and three thousand meters (possibly the highest seabird colonies on earth), have affected their reproductive cycle. They apparently initiate breeding a month later than they did a few decades ago, when they nested in lower, warmer areas. Hawaiian species are active on the breeding grounds only at night—most birds arrive in darkness, often only minutes after sunset.

Shearwaters and gadfly petrels breed in colonies and make use of a wide variety of breeding habitats in Hawaii: burrows, crevices deep in talus, dense vegetation. Several species use the same burrows at different times of the year, as they commonly do at Tristan da Cunha and the Salvages in the Atlantic. Most of these birds nest at sea level, but dark-rumps and Newell's fly to the interior of the main islands to raise their young in high mountain ridges and cliffs. Although wedge-tails in Hawaii are strictly coastal breeders, populations in the Kermadecs nest on mountain summits. Most shearwaters and gadfly petrels are hole or cavity nesters, and all successful nest sites provide incubating adults and young with protection from the hot tropical sun and predators. Inexperienced females sometimes lay in the open sun, but such eggs rarely hatch.

Wedge-tails, Newell's, dark-rumps, and Bonins may dig a new burrow each year or reclaim an old one. Most burrows have a single entrance. Bonins' burrows are the longest, measuring up to three meters in length and one meter in depth before opening into an enlarged nest chamber. Wedge-tail and Newell's burrows are somewhat shorter, one-half to two meters, and are usually gently sloped toward the nesting chamber. Most burrows are constructed with a bend that obscures incubating birds from external view. In the Northwestern Hawaiians, burrows in sand are typically dug near bunchgrass or shrubby vegetation that provides structural support. The sandy soil of Lisianski is so honeycombed with Bonin and wedge-tail burrows that human visitors constantly collapse the roofs of the tunnels. No doubt the length of burrows in each colony depends somewhat on local soil conditions. Dark-rumps dig substantial burrows wherever the soil in Haleakala Crater permits excavation; otherwise they use natural crevices or lava tubes for nest sites. Bonins, wedge-tails, and Newell's also adapt to local environments where the soil will not permit them to dig a burrow. Where cover is adequate, Bonins and wedge-tails sometimes nest on sandy surfaces or in shallow scrapes. Some wedge-tails nest in depressions on cliff slopes, deep ledges, and rock piles or in crevices on rocky islands such as Necker, Nihoa, Kaula, and Mokoli'i.

Neither Bulwer's petrels nor Christmas shearwaters dig substantial underground burrows. Bulwer's choose natural sites in rock rubble, crevices, caves, covered cliff ledges, or overhanging clumps of vegetation. They nest as far as possible from light. On low coralline islands such as Laysan, coral rubble provides shaded nest sites. Bulwer's are extremely opportunistic and readily nest in holes left by bombing activities or beneath buildings. They even have nested under dead turtle shells that a shipwrecked crew had heaped on a beach at French Frigate Shoals. Christmas shearwaters nest under dense vegetation such as beach magnolia or bunchgrass, in rock crevices, beneath buildings, and occasionally in abandoned wedge-tail or Bonin burrows. They will shape a tunnel through thick vegetation but will not dig in soil.

Courtship and Incubation

Like albatrosses, gadfly petrels and shearwaters tend to remain paired throughout their lives and have an uncanny ability to return to their nest sites after months at sea. Young birds return to the islands where they hatched after wandering the ocean for several years, and their ability to do so is based on environmental cues, not genetics. A young shearwater hatched from an egg that was transferred to another island returns to the island where it hatched, not to the one where the egg was laid. The first birds to return to the colony each season are those that already have mates. Males and females work together to renovate old, well-established burrows, spending about a week to clean out debris and loose soil before lining them with available vegetation. Bonins and wedge-tails begin new burrows by first loosening the soil with their bills. As excavation progresses, the birds lie on their sides to kick loose sand and soil backward in little jets. Bonins defend a three-meter radius around burrow entrances and repel intruders by dashing forward with outstretched wings and uttering loudly *kik-ooo-er*.

The return of Bonins to land after a few months' absence is awe-inspiring. Walter Rothschild wrote of their arrival on Laysan in August 1891:

> The next morning there were more, and on the third thousands filled the air. The new guests were pretty birds, barely the size of a domestic pigeon, but they began to domineer all over the island. . . . They are, on land, entirely nocturnal, and at once took possession of their innumerable subterranean burrows. In the bright moonshine one could see how they were busily engaged in removing the loose sand from holes, most of which had more or less collapsed since they had left them. Loving couples selected their nests and fought hard for them against later intruders. Quarrels, fights, and clamor became unceasing; in a few days there was no spot with sandy soil where the horrid 'song' of these petrels could not be heard. . . . The face of the island was entirely changed.

Virtually all of the obvious nocturnal activities in the colonies, including the complex aerial displays and the vocalizations so characteristic of this family, are connected with pair formation. Males' calls differ from females' and individual calls vary widely, so that birds can probably recognize each other by voice. Bonin and dark-rump males circle the colony chasing females, the pairs making short chattering cries to each other. Newell's call almost continuously when they circle their colony at dusk. Wedge-tails rarely call in flight, but pairs sit on the ground opposite each other, puffing up their throats and uttering prolonged two-part wailing duets that rise and fall in volume. Wedge-tails moan both inhaling and exhaling. Christmas shearwaters fill the air with groans similar to those of wedge-tails, but have three-note rather than two-note calls. Bulwer's remain silent in the air over the colony.

As night wears on, gadfly petrels settle onto the ground. The calls of most Hawaiian shearwaters and petrels subside after midnight, although intermittent cries, gurgles, and screams may last until dawn. Christmas shearwaters, in contrast, are most vocal around sunrise. They engage in little overt activity during daylight hours. They usually sit quietly side by side in the shade near their nest sites all day and are reluctant to take flight, scuttling under the scrub when they are disturbed. Christmas shearwaters make some low-altitude chases over the colony during the day, sometimes calling in flight. Wedge-tail pairs sit near their burrow entrances and keep up low, constant cooing much of the day while they rub each other's head and neck.

The Hawaiian names of the birds reflect their calls. The name *'a'o* (Newell's shearwater) mimics the call that to the ancient Hawaiians was thought to be an omen of death, perhaps because the bird sounds like a baby being strangled. More recently the biologist John Sincock has likened the voice to a combination of a jackass's bray and a crow's call. Dark-rumps and wedge-tails, known as *'ua'u* and *'ua'u kani*, have long-drawn-out *u-a-u* calls that suggest the wail of a lonesome cat. Bulwer's petrels, *owow* and *'ou*, make guttural barking sounds that alternate in a duetting pattern ad nauseam.

Shearwaters and gadfly petrels mate on land, usually near the nest site or within their burrow. Immediately after mating, pairs return to sea for about three weeks, to allow the females to feed while the eggs develop in the oviducts. During this prelaying exodus, nocturnal activity is remarkably subdued and the colony may seem deserted. Birds that land are usually displaying prebreeders. The return to sea marks the end of intense interaction between mated pairs. Theodore Simons estimates that parent dark-rumps spend only two or three hours together after the prelaying departure in early April.

Five species have synchronous laying periods of two or three weeks, during which most eggs are laid. Bonins tend to lay in late January, dark-rumps in early May, Bulwer's and Newell's in late May or early June, and wedge-tails in mid-June. Christmas shearwaters are less synchronous, laying from April to June. Females usually lay within a day of returning from sea. The single egg represents a high proportion of the female's body weight—from 18 percent for dark-rumps to 22 percent for Bonins and Bulwer's. The large egg and correspondingly large chick enables hatchlings to withstand temporary food shortages. Perhaps because of the physiological investment that a large egg represents, shearwaters and gadfly petrels do not replace lost eggs, even if an egg is lost immediately after it has been laid. Males and females of all Hawaiian species develop brood patches for incubation. Bonins and dark-rumps usually face toward the rear of the burrow when they incubate, perhaps to minimize the chance of attracting the attention of predators. On Midway, however, the strategy backfires. Rats enter Bonin burrows and steal eggs or small chicks from incubating birds before they are aware of danger.

Males and females alternate incubation bouts. Females incubate the initial day or two until the male arrives to take the first long incubation shift. The average length of a bout varies by species, from five days (Christmas) to twelve days (dark-rumps). During incubation adults do not eat, drink, or defecate. They minimize their energy requirements by spending virtually all of their time dozing, often with their bills tucked into their wings. They metabolize fats during their long fasts, producing water as a by-product. In contrast to tubenoses elsewhere, Hawaiian seabirds seldom neglect their eggs. On the contrary, dark-rumps and Bonins are particularly attentive to their eggs during incubation and rarely leave them for even a few minutes. Incubation periods are long in comparison with those of temperate seabirds but are fairly typical of tropical procellariids. The incubation period for all Hawaiian shearwaters is about 53 days. Gadfly petrel periods are more variable: 44 days (Bulwer's), 49 days (Bonins), and

55 days (dark-rumps). About five or six days before hatching, the chick begins to rub its egg tooth against the shell, which develops a star fracture. A few days later the chick begins to peep, alerting the parents that hatching is imminent.

Raising the Chick

Most chicks of each species hatch about the same time, usually in July or August. Christmas shearwaters tend to hatch during June and most Bonins in mid-March. Although chicks emerge wet from the eggshell, within a few hours they have become dry, round puffballs of grayish or blackish down. The down erupts in two phases. The light down is replaced after a few weeks by a second layer of heavier down.

Most chicks do not open their eyes for several days, but Bulwer's petrels come into the world with eyes open. Hatchlings are able to support themselves and can move about in the nest chamber after the first day, but spend most of their time sleeping. Parents brood young for the first few days, until they are capable of maintaining their body temperatures. Within a week both parents spend their time foraging for the nestling. Chick growth is adapted to meager, distant, or fluctuating food resources and accordingly fledging periods tend to be long and flexible. During good times chicks accumulate large fat reserves. Within a day of hatching, Bonin chicks receive meals that exceed half their body weight and dark-rumps can double their weight at a single feeding.

Parent dark-rumps initiate feeding within the first few hours of hatching by nibbling at the hatchling's bill, then regurgitating stomach oil and semidigested fish and squid. Wedge-tail parents prompt feeding by actually picking up the hatchling's bill. Later, whenever an adult returns to the nest chamber, dark-rump and wedge-tail chicks incessantly cheep and nibble at the parent's bill and throat to stimulate regurgitation. Most parental feeding visits are nocturnal and seldom last more than an hour.

Although feeding frequencies vary widely among colonies and from year to year, Hawaiian gadfly petrels and shearwaters are fed much less frequently than temperate species. Bonins feed their young every fifty hours; wedge-tails, Christmas, and Bulwer's are fed about every twenty-four hours. Dark-rumps' meals vary from 10 to 110 grams, the latter figure representing more than one-quarter of an adult's body weight; it is probably the maximum payload an adult can deliver. Weight gain is erratic. Chicks often gain for a week or so, then shrink when they receive no food for several weeks. Stomach oil, nine times richer in calories than fish, is helpful as a dietary reserve during starvation periods—chicks retain some in their stomachs even after they have eaten nothing for ten days.

Most breeding failures can be attributed to egg loss. As many as half of the eggs do not hatch, some because they are infertile. Other reasons for egg loss are as varied as the species and the breeding islands—predation by black rats (Midway), Polynesian rats (Kure), dogs (Kauai), mongooses (Oahu, Maui, Molokai, Hawaii), common mynahs (Kauai), Laysan finches (Laysan), Nihoa finches (Nihoa). At one rat-ridden colony on Midway, Bonins manage to hatch fewer than 10 percent of their eggs. Where disturbances are minimal, hatching success often exceeds 70 percent and can surpass 90 percent.

Fledging success varies widely by location. In predator-free colonies, about 85 percent of the Bulwer's, wedge-tails, and Christmas shearwater chicks that hatch survive to fledge. The reasons that young birds die are not always known, but most that perish do so during the first few weeks of life. Some die from exposure or because they are too weak to survive. Others starve because their parents cannot provide sufficient food. Before predator control programs were instituted, only about a quarter of the dark-rumps at Haleakala Crater fledged, primarily because the young were eaten by mongooses or feral cats. In predator-free areas, success rates approach those of other species. Rats eat young chicks on Midway and Kure, severely limiting fledging success. The rat problem on Kure was so serious in the mid-1960s that no Bonins fledged there for five years. After Warfarin was used to control rats, about half of the eggs that were laid produced young.

Some losses stem from interspecific competition for desirable nest sites. Bonin chicks are defenseless when adult wedge-tails return to reclaim and excavate their burrows on Laysan and Lisianski. Christmas shearwater chicks are sometimes killed when red-tailed tropicbirds or brown noddies take over their shaded nest sites. Weather also poses problems. High winds and surf associated with storms cause losses of Bonins and Bulwer's in the Northwestern Hawaiian Islands. In the main islands, heavy storms during breeding season can flood burrows and cause erosion. Severe rainfall in October has buried fledgling wedge-tails on Manana and Kauai and driven Newell's from their burrows. Fledglings can be attracted to lights on the main islands, with resultant disorientation or death.

All shearwaters and gadfly petrel nestlings are heavier than their parents. Bonins and Bulwer's typically weigh at least half again as much and some dark-rumps weigh twice as much as their parents. Dark-rumps receive about 70 percent of their food during the first half of the nestling period, when growth is concentrated on tissue and body parts. Later, tissues mature and flight feathers develop. Dark-rump chicks are usually fed only a few times during the month before they fledge, at which time their weight plummets, often dramatically. Desertion varies widely among species. Bulwer's petrels and Christmas shearwaters do not seem to desert their young in Hawaii, but other species apparently stop feeding the fledgling before it leaves the colony. Because feeding is sporadic toward the end of the nestling period, it can be impossible to determine whether parents have deserted a chick or whether they are simply foraging at sea for the next meal. Desertion varies within the same species. Some dark-rumps are deserted by their parents for up to six weeks before they fledge, while others are fed regularly until they depart.

Two months before fledging, chicks begin to exercise their wings with a series of rapid beats. A few weeks before departure, fledglings venture from their burrows, explore their immediate surroundings, and become conspicuous on the surface of the colony. The number of days required for fledging varies in proportion to the size of the bird. Fledging time also varies within each species with local feeding conditions. The larger birds—wedge-tails, Christmas, and dark-rumps—usually take 100 to 115 days. The medium-sized Bonins require 81 to 84 days and the small Bulwer's take 63 to 70 days. Fledging dates vary correspondingly. Bonins depart the breeding islands in late May, Bulwer's in September, dark-rumps in October, Christmas and wedge-tails in November. It is not clear what stimulates a chick to leave the colony, but departure may be prompted by hunger or weight loss. Parents do not teach their young to feed at sea, and each chick sallies forth into the Pacific armed with only its instincts and its accumulated fat deposits.

Conservation

The most serious conservation problem for gadfly petrels and shearwaters in Hawaii is the presence of alien predators in their colonies: mongooses, cats, pigs, dogs, rats, barn owls. This problem is particularly serious in the main islands, where dark-rumped petrel and Newell's shearwater populations are so low that the birds have been listed under the state and federal endangered species acts. Aboriginal Hawaiians contributed to the declines in the main islands. Young, fat nestlings are especially good eating and were reserved as delicacies for the *alii,* who feasted on them without apparent concern for any long-term consequences. Hawaiians easily pulled young dark-rumps to the surface by inserting long sticks into burrows and twisting them to spear downy birds. Such practices continued into the 1940s, perhaps later. Today mutton birds, which include several species of shearwaters, are harvested in Australia and New Zealand for food, sun-tan oil, and the enrichment of skimmed milk.

Dark-rumps once nested in the mountains of all of the main islands in countless thousands but now are restricted to the rim of Haleakala Crater, Maui, with small, remnant colonies on other main islands. They were once so common on Molokai that they were said to darken the sky of Pelekunu Valley. Pu'u 'U'au (dark-rumped petrel hill) on Oahu undoubtedly was once the site of an important dark-rump colony, and Alan Ziegler has found vast numbers of dark-rump bones in middens near Barber's Point. A cave above 3,000 meters on Mauna Kea, Hawaii, which was used as long ago as 1800, also contains bones. Today cats and mongooses take adults and young at higher elevations and barn owls do so in lower areas. Some fledgling dark-rumps may be confused by street lights on Maui when they first venture to sea, and those that alight on roadways may be run over by cars. Blood smears of dark-rumps found on Kauai in 1961 revealed serious cases of avian malaria, which may limit populations of dark-rumps and Newell's in the mosquito belt between the shoreline and the mountains.

Newell's were thought to be near extinction when a pig hunter in the Anahola Mountains, Kauai, located a colony in July 1967. Today the only confirmed colonies are on Kauai, although large colonies once inhabited the mountains of Maui, Molokai, and Hawaii (Waipio Valley). Kauai, Lanai, and Niihau are the only main islands that lack mongooses. Newell's were once so common near the town of Waiohinu, Hawaii, that a cliff was named *Pu-a-'a'o* (Pu'u Ha'ao on recent maps), or "flock of *'a'os*." During historical times, Newell's have suffered predation by pigs, cats, mongooses, black rats, and barn owls. Fires in dense uluhe fern destroys some colonies. A few cases of avian pox have been found among Newell's shearwaters on Kauai, probably contracted from alien birds. Like all gadfly petrels and shearwaters, Newell's are attracted to lights. Adults are sometimes dazzled when they fly by light sources and can be attracted to the running lights of ships. Newell's fledglings may be instinctively drawn to lights because they feed on bioluminescent prey. Their fatal attraction to lights is exacerbated by the fact that developing areas of Kauai are located between the colonies and the sea. As a result, disoriented birds crash and die in urbanized parts of Kauai each fall. Since 1968 a well-publicized program to recover disoriented fledglings and return them to the ocean has recovered some 20,000 young birds. Dead Newell's are also occasionally found near the Pali Tunnel, Oahu, especially during November.

Surveys to locate additional colonies of dark-rumps and Newell's on Kauai, Lanai, Hawaii, and Molokai would be welcome. Several dark-rump fledglings are usually recovered each fall on Kauai (Kapaa, Hanalei, and Waimea) during the Newell's shearwater recovery program, and since the early 1970s several have turned up on Maui. Calling dark-rumps may signify a colony near the Hono O Na Pali Natural Area Reserve on Kauai. Dark-rump fledglings also occasionally are found on Lanai, where colonies of forty to sixty pairs are suspected above 800 meters at Kumoa Gulch and Lanai Hale. Kilauea Crater, Hawaii, has yielded adults on eggs and a few dark-rump fledglings. Colonies are suspected along the Mauna Loa summit trail and on Mauna Kea above 3,000 meters near Pu'u Kanakaleonui. Newell's are frequently sighted on Hawaii, and two adults, one of them incubating an egg, was found in Hawaii Volcanoes National Park in 1972. Colonies are suspected at Makaopuhi Crater in the park and in the dense rain forest along the Hamakua coast and the Kohala Mountains, especially Waimanu Valley. Newell's may also nest in the Molokai Pali, especially near the rims of the Pelekunu and Wailau valleys.

Conservation problems for gadfly petrels and shearwaters in the Northwestern Hawaiian Islands are generally less severe than those in the main islands because the remote locations minimize disturbance by humans. Nevertheless, the actual or potential introduction of predators remains the most serious conservation problem. Black rats at Midway have eliminated a colony of 600 Bulwer's petrels on Eastern Island and have severely depleted the populations of Bonins, wedge-tails, and Christmas shearwaters on Sand and Eastern. Polynesian rats have had a similar effect on the seabirds on Kure. Midway had an effective rat-control program until at least 1970, but such efforts today are sporadic and often ineffective. Other Northwestern Hawaiian Island colonies are ratfree, but the concentration of 97 percent of the Bulwer's on Nihoa, three-quarters of the Bonins on Lisianski, and two-thirds of the wedge-tails on Laysan underscores the severe consequences that could be associated with an accidental grounding of a vessel and the introduction of rats or cats. A Japanese fishing vessel that went aground on Laysan in 1969 had evidence of rats on board, and it may be a minor miracle that none became established ashore. Logistical constraints in remote islands would complicate, if not prevent, effective rat-control programs. Elimination of rats would be extremely difficult on an island offshore Oahu, and would be both difficult and expensive on a Northwestern Hawaiian island.

Gadfly petrels and shearwaters are prone to indirect disturbance when their nesting habitats are disturbed. The devegetation of colonies, as by rabbits during the early twentieth century, or the introduction of exotic plant diseases could eliminate shaded nest sites for Christmas shearwaters and potentially undermine the soil in which Bonins and wedge-tails burrow. On Kure and Midway, colonies must be protected from conversion to other uses or landscaping projects that are not designed with the integrity of the nesting habitat in mind. The mere alteration of habitat by humans is not necessary deleterious—the largest Bonin colonies today on Sand Island, Midway, are on artificial hills near the harbor. The introduction of soil and vegetation on Sand has significantly improved the stability of burrows there.

Shearwaters and gadfly petrels face potential threats from the ocean. Dark-rump eggs have DDE residues of about 0.43 parts per million. Wedge-tail eggs contain low levels of mercury, endrin, DDE, and PCBs throughout the Hawaiian Islands. Such levels do not indicate immediate threats but do indicate a general exposure to toxic chemicals. Paul R. Sievert has found that about one-fourth of the stomach samples from Bonin petrels contain small plastic fibers, pellets, or fragments. Plastic is found to a lesser degree in all shearwaters and gadfly petrels in Hawaii and may be a source of PCBs in eggs. Plastic may result from misidentification by these omnivorous birds or from its presence in the fish and squid that they consume.

Certain fisheries could pose conservation problems for gadfly petrels and shearwaters. Because wedge-tails and Christmas shearwaters need predatory fish schools to feed, overfishing of tunas near colonies could adversely affect their ability to feed their young. In addition, goatfishes and mackerel scad, which are important food items for these shearwaters, have been proposed as baitfishes to support a tuna fishery. The high-seas drift gillnet fishery for squid north of Hawaii may trap and drown Hawaiian dark-rumps, but Japan, the Republic of Korea, and Taiwan are only beginning to provide scientific assessments of avian by-catch in that fishery.

11 STORM-PETRELS
Family Oceanitidae

Storm-petrels are so secretive that we know little of their habits. Yet the first storm-petrel that I encountered in Hawaii sought me out. Anchored a few kilometers offshore Lisianski on the fishing vessel *Easy Rider* in early May, I relaxed on deck quaffing an ice-cold beer after a simple evening meal. Cold beer is a rare luxury for field biologists in the Northwestern Hawaiian Islands and tastes especially good after twelve hot hours ashore slogging through the sandy terrain censusing seabirds and collecting food samples. The aft deck is the vessel's work space and the only pleasant spot aboard. As an interloper among five marine mammalogists who had arranged an emergency charter to investigate a massive die-off of endangered Hawaiian monk seals, I was relegated to sleeping space deep within the bowels of the ship. It took an acrobatic act to swing into my upper bunk, and my accomodations were so restricted that when I sat up in the night I cracked my skull on the bulkhead. Unless I used my headlamp, it was impossible to read there, and the stench from the malfunctioning head nearby prompted me to minimize my time belowdecks.

Sometime well after sunset I heard a light thud. Flapping on the deck in dazed confusion was a dainty, sooty-brown bird. Like storm-petrels throughout the world, this sooty storm-petrel was attracted by the deck lights of the ship and crashed its way aboard. Its wings had diagonal buff bars, its short bill was ebony, and because its tail was deeply forked rather than wedge-shaped, I immediately knew it could not be a Bulwer's petrel. As I picked up the trembling creature, it vomited a well-digested soup onto the deck, thus volunteering one of the few food samples I ever obtained from a storm-petrel. After a brief examination, I released the bird over the guardrail, and after a few flutters that encircled the ship, it disappeared into the inky night. This brief episode was the highlight of my day. We still do not know whether sooties breed on Lisianski, but I have found mummified bodies there. What was it doing just offshore Lisianski in the waning weeks of its breeding season if not at least prospecting for a nest site?

Storm-petrels are the smallest seabirds and inhabit all waters of the Pacific from Antarctica to the Aleutians. With hooked beaks and webbed feet, they appear to be diminutive versions of albatrosses, but their long wings are proportionately less narrow and their tube nose has a single opening. Most are dusky or gray and are so similar that they are exceedingly difficult to identify at sea except under exceptionally good viewing conditions. "Stormy petrels" were named from their habit of hiding during storms in the lee of ships, where they sought calmer waters to escape the worst effects of gales. They have long been part of the lore of the sea, each bird being said to be a reincarnation of a drowned person. It was of course bad luck to kill one. Also known as Mother Carey's chickens, they are named after the legendary female deity of the sea to whom ancient sailors called for divine help during foul weather. The appearance of the frail birds near a ship in distress was a sinister turn of events to sailors, who feared the birds were preparing to carry off the souls of the drowned. The genus of both species that breed in Hawaii is called *Oceanodroma*, Greek for "ocean runner," an apt name for birds that seemingly walk on water.

Modern taxonomy places the twenty storm-petrels of the world in eight genera, which can be divided roughly into Northern and Southern Hemisphere forms. The species that breed in the north, including all *Oceanodroma*, tend to have longer, more pointed wings and shorter legs than their southern relatives, and a ternlike flight. Some storm-petrels wander great distances, but most remain fairly localized. Sooty (Tristram's) storm-petrels are restricted to the North Pacific Ocean, with known breeding colonies in the Hawaiians, Volcanos (Iwos), and southern Izus, and possibly in the Bonins (Ogasawaras). Although the Hawaiian population probably numbers no more than 10,000 birds, it may account for the bulk of the world population. Sooties are known to breed only at Nihoa, Necker, French Frigate Shoals, Laysan, and Pearl and Hermes Reef, but may breed in small numbers on other Northwestern Hawaiian islands.

Harcourt's storm-petrels, also known as Madeiran, band-rumped, or Hawaiian storm-petrels, are birds of the northern tropics and subtropics. In the Pacific they breed in Japan (Hidejima), the Galápagos, and Hawaii. In the Atlantic they breed at Madeira and various other islands between St. Helena and the Azores. In Hawaii, Harcourt's probably breed on Kauai in the canyon walls of Hanapepe Valley and possibly in Waimea and Olokele canyons and on the Na Pali coast. The recovery of a few fledglings near Kilauea Crater suggests the existence of a breeding population on Hawaii. No nests or eggs have ever been found because the sheer terrain where they seem to nest renders the acquisition of information, including population estimates, virtually impossible. Downy fledglings have turned up on Kauai at Makaweli Beach and at the foot of inland cliffs where young birds have fallen after premature attempts to fly. A half century ago George C. Munro heard a Harcourt's storm-petrel squeak during the night in Hanapepe Valley and more recently John Sincock heard one at Hanapepe Lookout. A few were collected on Niihau in the late nineteenth century. Like young Newell's shearwaters, Harcourt's young seem to be attracted to street and hotel lights on Kauai during fall; one or two are recovered there most years and later released at sea.

Harcourt's storm-petrels are the smallest seabirds in Hawaii. At about 43 grams, their weight is less than half that of a Hawaiian sooty storm-petrel. Mostly dark brown, Harcourt's can be easily distinguished from sooties in the field by the presence of a broad white band that extends around the rump onto the flanks. Their tails are barely forked. Harcourt's fly with steady, rather high wingbeats interspersed with glides. Sooties, in contrast, have almost a swallow-like flight, fluttery wingbeats alternating with strong, steep-banked arcs and glides.

Like other tubenoses, storm-petrels tend to be long-lived and begin breeding at a relatively advanced age. The oldest sooty known is only about ten years old, but few banding studies have been possible because of the difficulties of gaining access to colonies during their winter breeding season. Other storm-petrel species live to fifteen or twenty years, and undoubtedly sooties and Harcourt's have similar life spans. While some storm-petrels breed as early as age three, Harcourt's first breed at five years in the Galápagos and seven years in the Salvages. As a consequence, as many as a third of the birds at sea are juveniles.

Storm-Petrels at Sea

Storm-petrels are perfectly at ease when raging winds hurl clouds of spray leeward from the tops of waves, filling the air with spindrift. During storms, much of their time is spent flying in slow wind eddies found in troughs, where, dwarfed by the crests of towering waves above them, they find shelter. In Hawaii, storm-petrels usually experience fairly calm, glassy surfaces or gently rolling swells, but sooties receive the full brunt of winter storms in the Northwestern Hawaiian Islands. Their wings maintain full control of the body during feeding movements, the head dipping to snatch up minute food items. Harcourt's hold their wings aloft in a V formation when they feed over concentrated food sources, like butterflies on a wet sandbar. Sooties briefly strike the water surface when they seize small food items but never submerge. On rare occasions sooties settle on the surface in large rafts, floating buoyantly. Storm-petrels are always close to the water, fluttering sporadically with their legs stretched downward, pattering with their webbed feet, lightly pushing off from the sea's surface. One problem with surface feeding is the danger of being eaten by predatory fishes. Many Harcourt's on Ascension sustain injuries to their legs, including the loss of toes or even entire feet.

All storm-petrels feed alone or in small groups and normally do not mix with other species except during temporary concentrations at a food source. One reason for a lack of sociability may be the habit of flying low over the water, which limits the view and may make it impossible to observe other birds

Sooty storm-petrel

feeding nearby. Hawaiian sooties and Harcourt's seem to do most of their foraging well offshore north of the islands, and local fishermen encounter the largest numbers of Harcourt's about five kilometers north of the Na Pali coast. Both storm-petrels disperse toward Japan after breeding, although some Harcourt's turn up south and southwest of the archipelago. Not surprisingly, both species are most commonly observed offshore near their colonies. Sooties often migrate in groups from their Northwestern Hawaiian Island colonies. Hawaiian storm-petrels seem to do most of their feeding at night, and the birds are probably attracted to food by scent or by light organs on the prey. The few Hawaiian food samples that have been studied indicate that sooties feed primarily on squid and midwater hatchetfish. They also take shrimp, isopods, amphipods, wind sailors (a coelenterate), and sea striders. All of their prey is quite small, and their pattering feeding habit permits sooties to pick dozens of minute objects from the surface in rapid succession. No food samples of Harcourt's in Hawaii have ever been analyzed, but elsewhere they eat crustaceans, fish, oily carrion, and garbage. Because the biomass of storm-petrels in Hawaii is so small, the food that they consume has little significance for the marine ecosystem.

Storm-Petrels on Land

The natural histories of most storm-petrels are poorly known. This is especially true in Hawaii, where both breeding species nest in locations that are inaccessible during all or part of the year. Consequently, some of our information comes from studies elsewhere. The breeding season of Hawaiian storm-petrels is timed to balance the availability of breeding sites with optimum feeding conditions. Sooties breed during winter throughout their range, apparently to avoid competition for nest sites with the larger Bulwer's petrels and wedge-tailed shearwaters, which reproduce in summer. On Laysan, adult Bulwer's and wedge-tails kill and forcibly remove sooty storm-petrel chicks that have not yet fledged when they reclaim their burrows in spring. Harcourt's breed in Hawaii and Japan during summer, when food supplies are probably at their maximum. Apparently other birds do not compete with them for small nest holes in the cliffs of Hanapepe Valley, although white-tailed tropicbirds conceivably could occupy large ones. In the Galápagos, Harcourt's have evolved a remarkable reproductive cycle in which they time-share nest sites. Two entirely distinct populations use identical nest cavities by breeding in alternating six-month periods. Sooties return to their Hawaiian colonies as early as October and begin to vocalize in their burrows by mid-November. Harcourt's probably return to Kauai and begin to nest in April.

Sooty storm-petrels select a variety of nest sites, depending on local availability. On Nihoa they nest in inaccessible recesses in talus slides. On Laysan they dig in the sandy loam beneath beach morning-glory, use cavities in guano piles, or burrow under bunchgrass. On Pearl and Hermes Reef, sooties burrow under clumps of bermuda grass or bunchgrass or use cavities under coral rubble. Although no Harcourt's nests have been discovered in Hawaii, elsewhere they select sites on steep slopes and cliff faces where crevices and platforms are flat enough to hold an egg.

Storm-petrels are perhaps more faithful to their nest chambers than they are to their mates. The return to the same crevice each year facilitates the reunion and preservation of each pair. Adults arrive ashore in fresh plumage, having molted at sea during the nonbreeding months. The male arrives first and locates, adopts, and maintains possession of the site. He purrs to attract his mate. Although storm-petrels are absent during the day, the night air above a colony is alive with birds that flit about just above the ground. Mates probably recognize each other by voice and by the musky scent that is typical of tubenoses. Most storm-petrels pass between their burrow and the ocean only under the cloak of darkness, probably an adaptive response to predation by gulls, owls, and similar enemies. Though such predators are uncommon or absent in Hawaii, sooties and Harcourt's are nocturnal. The full moon inhibits breeders and nonbreeders alike from visiting the colony.

Although their courtship behavior is poorly known in Hawaii, most storm-petrels have elaborate aerial rituals in which courting birds weave back and forth over the colony with crooning or whistling cries. Paired birds preen each other with gentle nibbles at the back of the neck. Much of such activity is probably attributable to juvenile birds, which may prospect at a colony for several years before breeding. Storm-petrels pair up and obtain a nest site before the first attempt to reproduce. Storm-petrels' calls, unlike those of shearwaters, are rarely obnoxious or intrusive. Although Harcourt's are rare today in Hawaii and may number in the mere hundreds, the fact that Hawaiians gave them names that mimic their calls, such as *oeoe* and *akeake*, seems to indicate that Harcourt's were once much more common there. Alan Ziegler has found huge numbers of Harcourt's bones in middens on Hawaii and in coastal sinkholes on Oahu, where they have apparently been consumed to local extinction.

Nestmaking is mostly haphazard. The birds use their bills and feet to remove any litter that has accumulated in the nest chamber during their absence, and they then may line the chamber with any vegetation that is available nearby. Copulation is rarely observed but probably takes place in the burrow. Harcourt's return to sea for three weeks before laying, a practice that allows the female to nourish the growing egg, which can account for almost one-quarter of her body weight. The peak egg-laying months for sooties are December and January. A single immaculately white egg is laid in the burrow, usually at night. Sooties' egg-laying is less synchronous than that of most Hawaiian tubenoses. Mark J. Rauzon and Sheila Conant witnessed every stage of nesting, from incubated eggs to fully feathered chicks, on Nihoa in February 1981. Harcourt's incubation bouts in Hawaii are probably similar to those at Ascension and the Galápagos, which last about six days. The male usually takes the first long shift and alternates with the female throughout the incubation period, which lasts from thirty-eight to forty-two days. Storm-petrel embryos can tolerate a lot of chilling. Elsewhere, eggs that were abandoned for several weeks during food shortages hatched when the parents resumed incubation.

Probably about half of the storm-petrel eggs in Hawaii hatch, a rate similar to the success rates of Harcourt's at Ascension and the Galápagos. Most losses stem from infertility or competition among storm-petrels for the best nest sites. Hatchling sooties are covered with soft, straggling black down. They are weak and helpless, barely able to raise their heads from the ground or move their bodies. Their eyes have no immediate use in their dark burrows and remain shut for the first week. Parents brood the chick during the first week until it can regulate its own temperature. Once a chick can be left alone, the parents work full-time at the task of feeding the growing nestling and only occasionally spend daylight hours in the nest cavity. Most chick loss occurs within ten days of hatching. Starvation is rare except when a parent has died, but disturbance by other birds causes many deaths. Hatchlings are initially fed a slurry of stomach oil and predigested prey about once a day, but the frequency tapers off to alternate days as they grow. Chicks may be fed either during the day or at night, but rarely between midnight and noon. Returning parents are greeted by a cheeping youngster that urges them to regurgitate by nibbling at the head and beak. The transfer of food is swift—parents usually return to sea within half an hour.

After the first month a second coat of down develops, and feathers follow rapidly. After six weeks or so, fledglings weigh a third or even one-half more than their parents. They lose the extra weight before fledging, when the energy demands of feather growth and exercise outstrip the rate at which parents can deliver food. Chicks are fed until they fledge; the "desertion" that is sometimes described is merely the absence of parents between feedings. Fledglings depart the nest alone at night and must be immediately self-reliant because their parents do not help them to feed at sea. Sooties fledge by mid-May and Harcourt's by October, taking fifty-eight to seventy-two days to mature. About two hatchlings in three survive to fledge, for an overall success rate of only about one chick per three eggs.

On the Galápagos and Ascension, overcrowding is a major cause of mortality among Harcourt's chicks. In Hawaii, competition for nest sites occurs only between other species and is rare among storm-petrels. Although masked boobies and great frigatebirds sometimes kill storm-petrels, the biggest mortality factor is introduced predators. Rats take sooties on Midway and Kure, and mongoose, cats, rats, dogs, owls, and pigs will devour or harm any Harcourt's that attempts to breed at an exposed site in the main islands.

Conservation

Storm-petrels may not have been common in Hawaii during the past few centuries—certainly the earliest Western naturalists did not report many—but Harcourt's are common in some Hawaiian middens. Because of their uncertain status, sooty storm-petrels have been designated a sensitive species by the U.S. Fish and Wildlife Service, and the State of Hawaii has listed Harcourt's as an endangered species.

The most serious threat to sooties is the potential introduction of rats or other predators to any of their primary breeding sites: Nihoa, Laysan, and Southeast Island, Pearl and Hermes Reef. It would be extremely expensive and time-consuming to remove rats from Nihoa or Laysan, and it may be technologically impossible to do so. Small ground-nesting birds lack a means to defend themselves against large rodents, and could face extermination if predators were introduced. The sooty population at Torishima Island has declined toward extinction two decades after the introduction of rats. Although the records are unclear, sooties probably bred on Midway before the introduction of black rats in 1943. During 1980–81 Gilbert Grant found adult sooties with enlarged testes on Sand Island, Midway, in December and February, and located some small storm-petrel burrows in an old housing area. Sooties may be attempting to recolonize Midway, but it is unlikely that they will be successful as long as the rat population remains unchecked. Polynesian rats on Kure adversely affect sooty nesting there.

Because Mark J. Rauzon has confirmed the existence of a large sooty storm-petrel colony on Nihoa, the population in the Northwestern Hawaiian Islands is greater than it was once thought to be. Nocturnal surveys are needed during the winter breeding season to determine the actual sizes of all colonies, especially those on Necker and Kure. To be effective, a thorough survey must include the tape-recording of calls to discover birds deep in burrows. A more accurate status assessment throughout the range of sooty storm-petrels would help determine what conservation measures, if any, are needed. Although their precise nest sites are unknown, Harcourt's probably breed in such inaccessible talus or mountaintop areas that rats cannot reach them. Efforts to locate Harcourt's nesting sites in the mountains of Kauai (such as Hanapepe Valley), Hawaii, and Maui should be encouraged.

Storm-petrels are susceptible to marine pollution. Tiny plastic pellets and fragments, the floating detritus of our industrial age which emanates from both North America and Asia, are found in one-third of the stomach samples from sooties in the Northwestern Hawaiian Islands. Storm-petrel stomach oil contains traces of fossil fuels whenever a marine oil spill occurs, and offers a potential means of monitoring pollution. Large-scale human activities near the Northwestern Hawaiian Island colonies, including the presence of large numbers of ships, could increase mortality among sooties because of their attraction to lights. Collisions between seabirds and ships create hazards for humans as well as birds. A small fishing vessel in Shelikof Strait, Alaska, was so deluged by parakeet auklets landing on the deck that the scuppers clogged. The captain and crew had to shovel them overboard, fearful that the ship might sink into the icy northern waters.

1. Adult Laysan albatross feeding chick; inset: chick

2. Rain-soaked Laysan albatross chick

3. Adult black-footed albatross

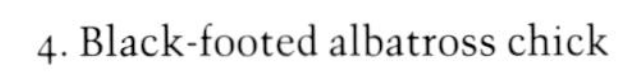

4. Black-footed albatross chick

5. Adult black-footed albatross with chick

6. Adult short-tailed albatross

7. Wedge-tailed shearwater chick

8. Adult wedge-tailed shearwater

9. Christmas shearwater chick

10. Adult Christmas shearwater

11. Adult Bonin petrel

12. Bonin petrel chick

13. Male great frigatebird

14. Great frigatebird pair

1. Adult Laysan albatross feeding chick; inset: chick

2. Rain-soaked Laysan albatross chick

3. Adult black-footed albatross

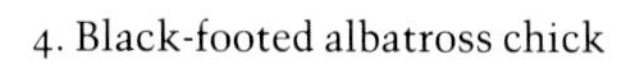

4. Black-footed albatross chick

5. Adult black-footed albatross with chick

6. Adult short-tailed albatross

7. Wedge-tailed shearwater chick

8. Adult wedge-tailed shearwater

9. Christmas shearwater chick

10. Adult Christmas shearwater

11. Adult Bonin petrel

12. Bonin petrel chick

13. Male great frigatebird

14. Great frigatebird pair

15. Great frigatebird chick

16. Great frigatebird chick

17. Adult masked booby feeding chick

18. Adult masked booby

19. Adult masked booby with chick

20. Adult masked booby

21. Masked booby pair

22. Adult red-footed booby

23. Red-footed booby

24. Adult red-footed booby

25. Red-tailed tropicbird chick

26. Adult red-tailed tropicbird

27. White tern chick

28. Adult white tern

29. White tern pair

30. Brown booby pair (♂, left; ♀, right) with chick

31. Sooty terns

32. Sooty tern colony; inset: adult feeding chick

33. Adult brown noddy

34. Adult blue-gray noddy

35. Black noddy chick

36. Adult black noddy

37. Gray-backed tern chick

38. Adult gray-backed tern

39. Laysan Island (photo by George H. Balasz, National Marine Fisheries Service)

40. Moku Hala, off Maui

41. Huelo Island, off Molokai

12 FRIGATEBIRDS
Family Fregatidae

Late on a blustery afternoon, Kilauea Point Lighthouse, Kauai, is an ideal spot to watch great frigatebirds. These spectacular creatures float high over the surging surf without the slightest effort on long, motionless wings, occasionally straying over the cliffs to eyeball any humans below. It is best to avoid standing directly under the sinister-looking birds—I have learned from sad experience that frigates are accurate bombardiers. Under ordinary sailing conditions the long tails are not noticably forked, but among the eddies and gusts of the cliff face frigates open and close them like a pair of shears to maintain balance.

When a red-footed booby flaps shoreward to its nest laden with a stomachful of flyingfish or mackerel scad, a soaring frigate may swoop into action. Skillfully flying circles around the hapless red-foot, the frigate pecks at the neck, tail, or wing to throw the heavier bird out of its flight pattern. The assault is not merely a series of pesky insults to torment and harass a neighbor. A fierce snap from a frigate's hooked bill may dislocate a booby's wing or cripple its leg, threatening its life. Although most attacks are unsuccessful, boobies stand little chance when a frigate is serious. It may dodge and croak its objections to such treatment but often disgorges its evening meal. The frigate seizes the booty in mid-air and may toss it skyward, where other frigates can pass the morsel back and forth a few times until one finally ends the fun by swallowing it. The Hawaiians appropriately named the great frigatebird *iwa*, or thief.

The prominent seabird biologist Robert C. Murphy described frigates as unparalleled flying machines. Their adaptation to life in the air is so perfect as to exclude any other mode of existence. Frigates' short, weak feet and legs are useless except for perching. Despite the unification of toes into a web, frigates never swim voluntarily. When they are grounded, the best they can do is shuffle along clumsily or climb to a perch high enough to permit a takeoff. A frigate's bones are light tubes with paper-thin walls, and the entire pneumatic skeleton makes up less than one-twentieth of its body weight. Almost half of the weight is accounted for by feathers and pectoral muscles. A frigate's breastbones (the pectoral girdle, consisting of the sternum, coracoids, and furcula), unlike those of birds in any other family, are fused together to provide additional rigidity and strength where the large flight muscles are attached. A frigate has a quarter more flight feathers and two-fifths more wing area than any other seabird of similar weight. This unique anatomy provides great dexterity and speed. The frigate can hover, soar, stall, twist, and glide with apparent ease, and its broad curving wings allow it to outmaneuver virtually any other bird. When searching for prey frigates are capable of sustained economical flight and flap only when the air is calm.

Christopher Columbus described the terrorization of other seabirds and the aerial recovery of regurgitated fish by these feathered buccaneers in the log of his voyage to discover America during the late fifteenth century; apparently he was the first Westerner to do so. The names frigatebird and man-o'-war hawk are derived from their habit of pursuing other seabirds. The swift-flying marauders were likened to the fast frigates or men-o'-war used by pirates on tropical seas. A single genus, *Fregata*, is recognized today; it includes all five species that inhabit the tropical and subtropical oceans of the world. Great frigatebirds are the only species that nests in Hawaii, and are inappropriately denominated *Fregata minor* because of certain peculiar rules employed by taxonomists. Lesser frigatebirds are occasionally sighted at Kure, French Frigate Shoals, and other Hawaiian islands, but no nests have been found despite the fact that lessers commonly nest in the Line Islands.

The long, slender wings, deeply forked tails, and saber-like bills render frigatebirds unmistakable in the field. Great frigatebirds are the most variable in size and color of the five frigate species. Hawaiian birds weigh much more than those in the Galápagos or on Aldabra or Christmas Island in the Pacific, and their wingspans exceed those of black-footed albatrosses. They are sexually dimorphic, adult females weighing almost one-third more than males. Males are entirely black above, with long scapular feathers that glisten with metallic green and purple. Their inflatable red gular pouches are used for mating displays on breeding territories. Soon after two birds have paired, the male's sac deflates and fades to a pale orange. Because male magnificent and great frigatebirds are so similar, some magnificents may be seen in Hawaii but are misidentified as greats. Adult females are also generally black but have white breast feathers and white or grayish skin on the throat. Juvenile plumage lasts between five and nine years and is extremely variable, tending to be primarily black with much white. The rufous feathers on juveniles' heads, necks, and breasts disappear after the first few years. Although it is almost impossible to identify any juvenile frigatebird at sea with certainty, those seen in Hawaii are likely to be great frigatebirds. Like other seabirds, great frigatebirds are long-lived. They live to be at least thirty years of age and some may reach forty or fifty.

Great frigatebirds range widely over the warm waters of the Indian and Pacific oceans but are restricted to Trinidad and Martin Vas in the tropical Atlantic. The world population is estimated at between one-half and one million birds. They breed on or visit all island groups in the tropical Pacific except Palau, ranging between 28 degrees north latitude (Kure) and 25 degrees south (Pitcairn).

While some colonies number tens of thousands of birds, no frigate colony in Hawaii approaches in size the colonies of sooty terns, wedge-tailed shearwaters, and Laysan albatrosses. Frigates nest on Kaula and on all of the Northwestern Hawaiian islands except Gardner Pinnacles. Nihoa and Laysan are the major Hawaiian colonies; only a handful breed near the main islands. The frigates commonly seen at Kilauea Point and along the coasts and in the valleys of Oahu roost offshore on Moku 'Ae'ae and Moku Manu. Although a few nest on Lehua and occasionally at Moku Manu, most of the thousand or so frigates observed in the main islands fledged elsewhere. Because frigates rarely begin to breed until they are at least nine years old, a substantial portion of the population consists of nonbreeders. Of the total Hawaiian population of 64,000, only about 20,000 birds breed. Although frigates have fairly large bodies, the population is so comparatively small that frigates consume far less than 1 percent of the marine resources consumed by the Hawaiian seabird community.

Frigatebirds at Sea

Although frigates obtain most of their food from the ocean, the majority are nonmigratory residents that do not wander far from their breeding islands. Residents return to land to roost on still evenings when air currents are unfavorable but otherwise may soar all night. Roosting frigates become active again just before daybreak, when they can ride updrafts above the islands so high that they become specks in the blue. At sea, great frigatebirds are most abundant during winter, when their numbers ashore decline and some postbreeding adults wander from their Hawaiian colonies as far as Asia and Micronesia.

Once a juvenile is capable of independent existence, it may become highly nomadic. Fledglings banded at French Frigate Shoals, Laysan, Pearl and Hermes Reef, and Kure have been recovered in various parts of the Philippines. Frigates from French Frigate Shoals and Kure have turned up at Eniwetok and Ugelang Atoll, Marshall Islands. More typically, adults and young move among the Northwestern Hawaiians and Johnston Atoll. Although most juveniles eventually return to the island of their birth, sudden increases in breeding populations, such as the dramatic recolonization of Baker Island by lesser frigatebirds after cats were removed, are probably due to influxes of nomadic young birds.

F. C. Hadden observed a remarkable interisland movement of great frigatebirds at Midway on December 29, 1938:

> A large formation of frigates appeared above Midway. They were coming from the southeast and flying directly into the wind which came from the northwest. Kure or Ocean Island is the only island to the northwest of Midway—they must have been headed for it. They formed a line 3 to 5 abreast for as far as the eye could see in both directions. Five hundred birds were counted, but some had already passed, and they were still coming when the observation discontinued. They flew into the wind at a height of about 2,000 feet without any apparent movement of the wings or feathers.

Great frigatebirds are pelagic feeders but usually stay within eighty kilometers or so of their breeding or roosting islands. They are usually solitary at sea, yet when ephemeral food is available they will join fairly large flocks of other species, especially sooty terns and wedge-tailed shearwaters. Frigates fly high over the ocean in search of food and can descend rapidly when other species locate fish. In mixed feeding flocks they soar to 160 meters, substantially higher than any other seabird. As the periods between surfacing by tunas or porpoises lengthen, frigates tend to soar higher and disperse. Frigates have highly specialized feeding methods—they use their long, hooked bills to snatch prey on the wing or during brief contact with the water. Because of a structural inability to take off again if they settle on the water (*not* a lack of oil glands and associated waterlogging), frigates are limited to snatching prey from surface waters, probably no deeper than the length of a bill. They never roost on water.

Frigates' diet in Hawaii includes about six-sevenths fish and one-seventh flying squid, together with the occasional juvenile sooty tern. Although crustaceans are occasionally found in frigates' stomachs elsewhere, none have turned up in Hawaiian samples. Frigates take fish of a particularly wide range of sizes, from a thumbnail-size cowfish to a halfbeak longer than this page. Flyingfish are the most common family of fish taken, especially *Cypselurus* species and Linne's flyingfish. Mackerel scad are frequently eaten during summer and fall. Some prey are taken primarily at certain locations or during certain months. For example, frigates consume many Pacific sauries near Midway during winter and numerous small fantail filefish near Laysan during summer, but eat neither of them in other areas or during other seasons. Frigates survive by exploiting any organism that happens to be locally available in surface waters, often flyingfish and flying squid. Although food studies suggest that sooty tern chicks are not commonly eaten in Hawaii, frigates can wreak havoc at some tern colonies. Hundreds of sooties are eaten each year on Tern Island, French Frigate Shoals, principally by females and juveniles. Frigates also may eat hatchling green sea turtles, gray-backed terns, fledgling shearwaters, and other young birds.

Early writers emphasized kleptoparasitism as frigatebirds' sole method of feeding. In fact, frigates in Hawaii earn most of their living by honest fishing; piracy is primarily a supplementary activity of females or of juveniles that are struggling to learn a highly specialized type of aerial feeding. Frigates' diets are quite different from those of the birds they typically harass at Hawaii colonies: red-footed boobies, red-tailed tropicbirds, wedge-tailed shearwaters. The frigates at Nihoa and Laysan are at least twice as numerous as the combined populations of all three booby species, and the boobies could not possibly survive such a large number of avian parasites. Theft may be most important during periods of food stress, when the ability to rob other species would enhance survival.

Frigatebirds on Land

Frigates are seldom endearing at close range. The effluvia emanating from excreta and rotting fish at a colony can be nauseating. When an interloper draws near, frigate chicks stare with a malevolent, insulting air and strike viciously with their long beaks. Too close an approach may entice one of the flat-bodied hippoboscid flies that infest this species to explore the hair of the intruder. Although the shortcomings of frigates can be ignored when they are chasing across the skies, ashore the reminders are constant that these winged cannibals engage in wholesale slaughter of each other's offspring. Marauding juveniles carry off and swallow unguarded eggs or hatchlings whenever an opportunity arises. On Laysan, adults sometimes toy with downy frigate chicks in the air, tossing them from beak to beak until one swallows the hapless nestling. Dismayed by such seemingly pointless behavior, J. Bryan Nelson wrote: "Frigates seem beautifully adapted for sabotaging their own breeding effort."

At rest, frigates often perch with wings outspread and turned upside down by forward rotation, a stance that aids cooling in the islands' breezes. When roused from their roosts by humans, displaying males sprawl over the bushes awkwardly. Frigates are generally silent, but when landing, displaying, fighting, or chasing they emit a confusing variety of clapping, gakkering, squealing, and warbling sounds.

Breeding Season and Nest Sites

Great frigatebirds in Hawaii breed during spring and summer; the peak egg-laying period usually comes between March and April. Most juveniles can fly independently by October. The nesting cycle begins when males roost atop vegetation in groups of fifteen to twenty birds at traditional nest sites. They inflate their distensible crimson gular pouches in an attempt to attract a mate. Because frigates perch so well, they use a wider range of nesting habitat than other arboreal seabirds, yet need at least a semblance of vegetation on which they can construct a nest platform. They are capable of taking flight in calm conditions provided sufficient air space is available. Landing is more difficult—nest sites are usually chosen to allow an approach into the prevailing wind. On the Northwestern Hawaiian Islands where beach magnolia grows, nests are built between one-half meter and four meters from the ground. On islands without shrubbery, frigates make use of plants so low that their nests may rise only a few centimeters off the ground. Frigates on Pearl and Hermes Reef construct nests on nightshade thirty centimeters or less from the ground. On the upper ridges and higher slopes of Nihoa and Necker they build nests on goosefoot, ohai, and ilima. On Whale-Skate Island, French Frigate Shoals, frigates get by with a foundation of puncture vines.

Courtship and Incubation

Because frigates cannot breed every year, the typical web of pair formation and territoriality in seabirds is altered substantially. Fidelity to nest sites and mates is impossible. Unlike many Hawaiian seabirds, frigates do not invest a season in occupying a particular nest site before they breed. Males are so loosely tied to their display sites that when one is unsuccessful in attracting a female, he simply flies off to another spot to join in or initiate a new group of displaying males. Because frigates breed at a new site at each nesting attempt, they devote little effort to site maintenance. Without a site to which mates return each spring, monogamy becomes problematical. In comparison with cold-water pelecaniforms, frigates spend little time or effort defending a nest site.

Displaying can begin in late December, peaks from late January through April, and usually ceases by May. Attracting a female can be very stressful. Males vanely posture in groups for up to three days without a break, bills pointed skyward. The scarlet balloon obscures the front of each bird so that only his bill and eyes peek over the top of the sac. Females register their interest by hovering overhead on virtually motionless wings while the male below whinnies and raises his vibrating wings. In 1906 Walter K. Fisher colorfully described the male's behavior when a female appears:

> He suddenly arouses himself from the lethargy, and as she passes he rises partially from a sitting posture, throws back his head, spreads his wings, and protruding the brilliant pouch, shakes his head from side to side, uttering a hoarse cackle. Occasionally, when the female alights near, he waves his pouch from side to side, the head being thrown well back and the wings partially spread. At the same time the long, greenish iridescent scapular feathers are fluffed up and the creature presents a most unusual and absurd appearance. In this posture he chuckles again and again, and rubs his pouch against his mate, who usually ignores him completely and flies away.

More recently, J. Bryan Nelson has graphically described their behavior:

> The wing-trembling is striking, for the males swivel on their perches as they orient their display directly toward the female, giving the effect of two enormous, beseeching black arms with the crimson sac couched between. If

> the female descends, the male's head-waving becomes more exaggerated and disjointed. A fine-drawn reeling sound and mandible-clapping, sometimes interspersed with a peculiar quavering sound, accompanies the display. Any one of three such conspicuous components would be dramatic; they produce an astounding performance when delivered by a group of birds.

Although most displays are directed toward females, males do defend a small territory, lunging and snapping at incoming males who attempt to land too close. Nevertheless, some pairs nest within touching distance.

A female indicates her choice by settling beside her favorite displaying male. Soon thereafter they copulate and build a nest at the display site. In some colonies nests are so dense that displaying males dislodge the eggs and young of neighbors with their wing movements. Disruption is minimized by the synchronization of breeding within each village of frigates due to the social stimulation from an abundance of displaying males. The male gathers nest material on the wing by picking up loose material from the ground, breaking off twigs from bushes, or robbing nests of neighboring frigates and red-footed boobies. The female must remain at the nest site to guard the flimsy platform of sticks, feathers, and other debris from marauding birds. Over the course of the season, the flat platform gradually becomes cemented together into a yellow, excreta-hardened pad.

Because most pairs remain together for but a single breeding attempt, pair-bonding rituals are minimal and last only three or four days. Frigates do not engage in courtship feeding. After the initial encounter, mating, and nest building, the birds have no further pair-bonding behavior. Unlike albatrosses and shearwaters, female frigates do not have a honeymoon at sea during which they feed while the egg develops. This hardship is minimized by the fact that a frigate's egg weighs only about one-twentieth of her body weight, a smaller proportion than that of any other Hawaiian seabird except boobies.

Typically only a single egg is produced, but a nest with apparent twins was found on Nihoa. Both sexes develop median brood patches and males usually take the first incubation shift. During the fifty-five-day incubation period, great frigatebirds in Hawaii change over about every three days. Great frigates in the Galápagos, by contrast, change every ten days, and those at Aldabra about once a week. Frigates have no special nest-relief ceremony: the incubating bird simply vacates the nest and its mate takes over. Platforms are so flimsy that eggs frequently roll off during change-overs or when neighboring birds take flight. It is not yet known whether great frigates replace lost eggs, but the four other frigate species can do so within two weeks.

Raising the Chick

Slightly over half of the great frigatebird eggs laid in Hawaii hatch, much more than the one-third or so that hatch elsewhere. Reliable measures of hatching success are very difficult to obtain because frigates are so susceptible to disturbance that the activities of well-meaning biologists may bias the information they obtain. A frigate hatchling emerges from the egg naked with bluish skin, bill, and legs and a head that seems much too large for its ugly little body. Although its eyes open shortly after hatching, the young bird is uncoordinated and lies helpless for days. Once it becomes alert, the frigate supports itself on the nest by clinging to sticks with its prehensile feet and long claws. The parents brood the hatchling closely to prevent the naked skin from being exposed to the sun and to protect it from being devoured by a juvenile frigate. Unattended small birds are usually attacked, evicted from their nests, and killed.

After about three weeks a coat of immaculate down transforms the wrinkled hatchling into a potbellied fluff ball. The growing bird can regulate its own body temperature within a month, and then both parents are freed to search for food. On Laysan, young frigates are fed every eighteen hours or so. When a parent returns to the nest, the nestling begs in a hunched posture with half-open wings, bobbing its head and shoulders and calling plaintively. The chick reaches deep into the throat of its parent, triggering the regurgitation of rotting fish and squid. After two months nestlings appear sinister—dense down covers the body and the back develops a black cape of scapular feathers. By the third month females are noticeably larger than males.

Nestlings can survive food shortages by growing slowly. They take between 120 and 145 days to attempt their first flight and achieve a measure of independence. Precocious young birds begin to fly in September and the laggards fledge by late November. Like no other seabird, frigates remain dependent on their parents for food for long periods of time after they are able to fly. During this period the fledglings attain the proficiency in accurate flying and specialized feeding which they need to survive. Juveniles play games by passing sticks to and fro or using flapping flags on ships as objects of sport. Such activities enhance and improve their coordination. Postfledging feeding continues as long as eighteen months in some colonies and commonly for a year. Typically, mothers feed juveniles much more often and much longer than fathers. The extra attention represents a huge investment in the young by the parents and renders annual breeding impossible for females. It is uncertain whether males attempt to breed each year, but if they do, they must find a different mate and probably another nest site.

Frigatebirds have poor success in raising their young. Their nesting success is much lower elsewhere than in Hawaii; often they need to lay seven eggs to achieve a single fledgling. One important cause of mortality is the disturbance caused by other frigatebirds at nest sites. Eggs and young can easily be dislodged from nesting platforms whenever unmated males attempt to set up breeding territories within a colony or when nesting material is pilfered. Disturbance by other frigates is exacerbated when humans enter a colony; such intrusions flush parents from eggs and hatchlings and create raiding opportunities for neighbor-

Great frigatebirds

ing frigates. Birds nesting early or late in the season or in areas of low density usually have better success rates because they have fewer neighbors to interfere with their reproductive efforts.

Chicks are most vulnerable during the first two or three weeks after hatching. Adults lower their body temperatures by soaring or stretching their feathered wings to catch the tradewinds. Because a hatchling cannot take to the skies and because spreading its puny wings would provide little cooling, exposure to sun can be fatal. Starvation claims some nestlings, especially the youngest. Polynesian and black rats may take eggs and hatchlings at Kure and Eastern Island, Midway, but conclusive information is absent. Two-thirds of the Hawaiian frigates that survive the first few weeks fledge, so that about one-third of the eggs laid in Hawaii yield flying young frigates. In the Galápagos, great frigatebird fledglings suffer huge drops in body weight, despite the feeding subsidies provided by parents, and most starve either during the long period of dependence or soon after the parents cease to feed them. Hawaiian frigates seem to fare better. Adults cut their losses and do *not* lose weight even when their young are starving. Evolution does not reward a long-lived bird for risking its life in a single attempt at reproduction.

Conservation

Probably the most significant conservation problem for Hawaiian great frigatebirds is the loss of nesting vegetation, as on Laysan and Lisianski early in the twentieth century when rabbits were introduced. The frigate population on Laysan may still not have attained its former level because the beach magnolia there has never competely recovered. Frigates are extremely opportunistic in fashioning nest sites, as evidenced by their ability to construct nests on low-lying atolls such as Pearl and Hermes Reef and French Frigate Shoals. Perhaps some would survive extensive devegetation.

Frigates are large enough to dispatch most terrestrial predators. They have survived the presence of rats on Kure and Midway, although such predators undoubtedly diminish their reproductive efforts. Frigates suffered virtually a total nesting failure on Kure in 1966, when many adults were attacked by Polynesian rats and received wounds on their backs. Frigates on Eastern Island nest adjacent to red-footed boobies, many of which are injured by black rats. The nesting distribution of great frigatebirds on Aldabra can best be explained as an attempt to avoid the local populations of black rats. Ralph W. Schreiber saw alien cats on Christmas Island completely destroy a nesting colony, eating all frigate eggs and chicks. An introduction of cats on any of the Hawaiian breeding islands would have to be dealt with immediately by conservation officials and land managers. Frigates were collected and salted down for food in the Bahamas at the turn of the twentieth century, but I know of no records of human consumption in Hawaii.

Frigates are particularly disturbed by human intruders in their colonies. Visitors, including biologists, should refrain from coming so close to nesting birds that they adandon their nests. Frigates once nested on Sand Island, Midway, but have not done so for decades, no doubt because of intense and regular disturbance there. Uncontrolled human activities not only can cause reproductive failure in the year of disturbance but can result in the abandonment of an entire colony. The British biologist A. W. Diamond observed the destruction of a great frigatebird colony at Aldabra when tourists from the M.S. *Lindblad Explorer* wreaked havoc on the birds. Fortunately, the colony was reestablished some years later. Magnificent frigatebird colonies in the islands and remote areas of Baja California are increasingly disturbed by photographers and amateur naturalists who do not intend the destruction that follows in the wake of their "innocent" trespass.

Frigates are capable of being tamed after capture, although their tempers cannot be trusted. After having been fed for a short time in captivity, birds in the Caribbean are content to sit on a perch and wait for meals to be handed to them. In Micronesia and Polynesia, frigates have been domesticated for use as interisland carrier pigeons. Relying on the instinct of frigates to return to their nests, Samoans used to erect perches near their homes and brought five-month-old chicks there to be fed. On foraging excursions the birds would roost on similar perches on other islands. Soon they were providing an ocean postal service for the transportation of small objects such as letters and shell fishhooks. Even today, trained frigates are sometimes taken on long canoe trips. A message placed in a reed cylinder attached to the leg or wing of the bird can be recovered a hundred kilometers away after the bird has returned to its perch.

13 BOOBIES
Family Sulidae

Late one May afternoon I discovered a cluster of six brown booby nests, a rare find on Laysan. The birds, perched tight like guardian gargoyles on nests of sticks and bunchgrass, were either incubating eggs or brooding hatchlings. When I approached, several began to barf up flyingfish and needlefish, thus volunteering samples for my study of their food habits. As I bent to retrieve the fish and place them in my collection jars, one female cried hysterically and, now lightened by the discarded ballast, bolted toward the sea. Her nest revealed two naked nestlings, one a feeble hatchling still sitting on its shell. The larger, perhaps a week old, immediately began to shove the smaller from the nest with its head. During the next few minutes, the elder pushed with its feet and pulled with its bill, gripping the neck and wing of its sibling until the tiny bird was beyond the perimeter of the nest. It would perish before sunset. I knew that fratricide is common when both booby eggs hatch and it seemed pointless for me to interfere with an event guided by millions of years of evolution. With my day pack bulging with a full load of samples, I trudged back to camp, brooding on the darker side of the struggle for existence in a tropical seabird colony.

Boobies are streamlined, torpedo-shaped birds with long, narrow wings. These goose-sized seabirds use their wedge tails to increase maneuverability when they wheel above the sea, fold their wings, and plummet to capture shoaling prey. The family Sulidae originated in the tropical Pacific Ocean and includes nine living species, all in the genus *Sula*. Fossil sulids date from at least 20 million years ago, at a time when Pearl and Hermes Reef was still a high island. Three species, called gannets, are now found in the temperate waters of the North Atlantic, Australia, and South Africa; they weigh about twice as much as boobies. The six boobies are tropical and typically sport brightly colored bare skin on their beaks, feet, and faces.

The term *Sula* is derived from an old Scandinavian word for Solan goose, or gannet. Spanish sailors believed they looked or acted stupid and called them *bobo*, meaning "dunce" or "clown." Early mariners plying tropical seas found they could grab boobies with their hands when the birds tamely settled down in the ship's rigging or on a boom. Any bird that alighted on a sailing ship became fresh meat for crewmen far from terra firma. Boobies' breeding behavior reinforces the pejorative term. Hawaiians call all boobies *a* (pronounced *ahhh*), apparently attempting to mimic their monotonous cries.

The three boobies that breed in Hawaii are infested with flat hippoboscid flies that thrive in the throat feathers. Masked boobies, also called blue-faced or white boobies, are the largest and most robust of all boobies, weighing more than two kilograms. As with all Hawaiian boobies, females weigh at least a tenth more than males, but large males weigh more than small females. Hawaiian boobies are substantially larger than those in the Galápagos, Christmas Island (Pacific), Ascension, and the Seychelles.

Adult male and female masked boobies have virtually identical plumages. Feathers are mostly a dazzling white, with brown-black tails and black trailing edges on the wings. The bare, blue-black facial skin forms a dark mask around the bill. The conical, serrated bill is straw-colored and the feet are yellowish gray. The two sexes look similar but during breeding season the males' bills are a richer, brighter yellow than the females'. The sexes are distinguished easily in the field by their voices because of anatomical differences in the syrinx. Females and all immatures have clamorous, low-pitched, guttural honks and croaks. Males sound as if they suffer from laryngitis, emitting various high-pitched, rather pathetic asthmatic whistles and hisses. From a distance, masked boobies in flight may be distinguished from similar red-foots by the presence of more black on the wings and tails and a heavier bill. Young masked boobies vaguely resemble brown boobies, having a dark-gray head and neck and generally grayish-brown upper parts. Second-year birds are a mottled gray-brown with a white head. All Hawaiian boobies attain adult plumage by their third year.

Brown boobies are handsome birds. The dark chocolate brown of the head, back, and lower neck contrasts sharply with the immaculate white of the breast and belly. Underwing linings are primarily white with brown borders. Although plumages are identical, the sexes can be distinguished at close range during breeding season: a male has a bluish and a female a yellow-green cast to the skin on the feet, face, and bill. The voices differ somewhat but not nearly to the extent of those of masked boobies. A juvenile's plumage is similar to that of an adult except for the lack of sharp color demarcation on the breast. A few hybrids of masked and brown boobies may be seen.

Red-footed boobies are the smallest birds in the family. The adult plumage is mostly creamy white, with trailing black edges on the wings and a yellow wash on the head and neck. As their name implies, their legs and feet are a coral red. Red-foots have a finely tapered blue bill with a pink base. Although ashy-brown and intermediate plumages occur elsewhere, they are rare in Hawaii. George C. Munro estimated that no more than one bird in fifty was a brown morph on Moku Manu, and I have found them to be even less common in the Northwestern Hawaiian Islands. The sexes are difficult to distinguish in the field. Females have lower-pitched voices than males, but it takes a lot of experience to discern the difference reliably.

Boobies tend to live fairly long lives and defer breeding until they have gained sufficient experience. Masked and red-footed boobies live to at least twenty-five and twenty-two years, respectively. The oldest known brown booby in Hawaii lived sixteen years, certainly less than maximum longevity. The major brown booby colonies on Kaula and Nihoa are difficult to visit, but it is reasonable to assume that some individuals of all species live more than thirty years.

Distribution and Abundance

Masked, brown, and red-footed boobies are widespread throughout the subtropical and tropical waters of the Atlantic, Indian, and Pacific oceans and breed throughout their range. Their world populations are unknown, but masked and red-foots are particularly abundant. Brown boobies are sometimes called "common" boobies, but though their colonies are numerous, populations in many areas are small. Clipperton has a colony of 15,000 birds; other large brown booby colonies are located in the Caribbean and the Gulf of California. In the Pacific, boobies breed on most island groups, especially where human disturbance is minimal. They are absent from or very rare in parts of the southwest Pacific, including the New Hebrides and the Bismarck archipelago.

In Hawaii, all three species breed on Lehua, Kaula, and most of the Northwestern Hawaiian islands, where they are year-round residents (Tables 1 and 2). The importance of each island varies with the species. Masked boobies number over 7,000 birds, a quarter of which are nonbreeders. Large colonies are located on Laysan, Lisianski, and French Frigate Shoals. The Hawaiian population of red-foots exceeds 23,000, almost half of which do not breed. They are most numerous on Nihoa, Necker, and Laysan. With fewer than 3,000 adults, only 2,000 of which breed, brown boobies are the least common booby in Hawaii. Large colonies are restricted to the rocky islands of Kaula and Nihoa. Brown boobies no longer breed on Midway and only a handful of masked boobies nest there, probably because of disturbance by humans.

Red-foots are the only boobies that breed widely in the main islands, with colonies that can exceed 500 pairs on Oahu (Ulupa'u Head, Moku Manu, Sea Life Park), Kauai (Kilauea Point), and Lehua. A few masked boobies breed on Moku Manu and Lehua at least during some years. They sometimes roost but apparently do not nest at Kawaihoa Point, Niihau. Brown boobies nest on Lehua and Moku Manu and roost on Moku 'Ae'ae (offshore Kilauea Point), the Na Pali coast, and Ki'ei Islet (offshore Lanai) and near sewage outlets or on buoys off

Sand Island, Oahu. Browns nested on precipitous cliffs on Niihau a half century ago, but I can find no recent confirmation of their presence there.

Because boobies breed younger than most Hawaiian seabirds, breeding boobies (20,000) outnumber immatures (13,000). The fact that red-foot populations have surged in recent years may account for the much higher percentage of nonbreeding birds among red-foots than among brown and mask boobies. Although boobies are fairly large, their populations are too small to have a great effect on the Hawaiian marine ecosystem. They consume about 2,500 metric tons of prey each year, less than 1 percent of the marine resources consumed by the Hawaiian seabird community.

Boobies at Sea

Boobies often escort or circle ships. They rarely trail a vessel but instead prefer to coast back and forth over the bow, hang-gliding with occasional wingbeats while taking advantage of wind deflected from the ship's bow. Flyingfish and halfbeaks are sometimes flushed by an oncoming vessel, and boobies are quick to seize in mid-air any organism foolish enough to break cover. Boobies will rest on almost any object of suitable size, including oil platforms in the Gulf of Mexico, floating debris in equatorial currents, and pitching buoys at Midway. Masked boobies sometimes ride turtles far at sea, sleeping with their heads tucked beneath their wings. Brown boobies often fish from perches on rocks, pilings, and docks.

Boobies are designed for plunge-diving and can make spectacular headlong hurtles into the sea from heights of up to thirty meters. During a dive the booby maintains complete control and is capable of aborting the effort a mere wing's length from the sea's surface. Its nostrils are sealed and air sacs beneath the skin of its throat cushion the impact when the bird strikes the water. The air sac is buoyant and limits the depth of a plunge to about five meters, but fully webbed feet aid the underwater pursuit of fish and squid. If a dive has been a success, the bird floats for a moment with its beak and face submerged beneath the water while preparing to swallow the catch, perhaps to minimize piracy by a frigatebird. Sometimes a booby reappears on the surface with a wriggling fish sideways in its bill; after some juggling, the bird swallows the fish. Boobies do not usually spear or impale fish, but I have found mackerel scads and flyingfish with wounds that could have come only from a booby's sharp beak.

As adult red-footed and masked boobies tend to remain close to the colonies, most birds observed far offshore are immatures or subadults. Young birds are highly nomadic and capable of interisland movements of thousands of kilometers. Movement among the colonies of the Hawaiian archipelago is common, and occasional birds banded there have been known to wander great distances. One red-foot banded as a nestling at French Frigate Shoals turned up at Wake and later bred on Kure. Red-foots have moved between Laysan and the Marshalls and between Wake and Lisianski. Browns banded on Kure have journeyed to Wake, Indonesia, the Marshalls, and the Tuvalus (Ellices). Masked boobies apparently limit their wanderings to the Hawaiian Islands. Thousands of Hawaiian red-foots, especially immatures, migrate after each breeding season to Johnston Atoll, where they roost conspicuously on the LORAN C tower and its support wires. By March, adults have returned to their colonies to begin a new season, and only juveniles remain.

Masked and red-footed boobies are pelagic feeders. Masked boobies range more than 150 kilometers from land, but usually return in the late afternoon to roost for the night, although some birds on Laysan arrive as late as midnight. Red-foots have a faster, more graceful flight than other boobies. They seek deep blue water for feeding, which at locations such as French Frigate Shoals can be just offshore. No booby in Hawaii regularly flocks with its own species during feeding activities, but masked and red-foots are moderately social at sea and sometimes forage with flocks of terns and shearwaters over schools of skipjack and yellowfin tuna.

All boobies feed by plunge-diving, using either perpendicular power dives or shallow dives at oblique angles. Brown boobies can dive repeatedly from heights scarcely equal to a wing's length, using low trajectories. The use of frequent shallow dives enables them to exploit inshore schools of small fish. They are strictly solitary feeders and tend to forage within sight of land. Brown boobies will accept a free lunch when one is available. In Venezuela and the Galápagos, they are known to congregate with shearwaters, frigatebirds, and storm-petrels to feed on fish that float to the surface when they become incapacitated by sulfurous fumes emanating from volcanic activity or fissures in the ocean bottom.

All Hawaiian boobies consume many adult mackerel scads and flyingfish (especially Cuvier's, Linne's, and short-winged flyingfish). Large flyingfish, some more than a third of a meter long, dominate the diet of masked boobies and account for almost three-fifths of their prey. Masked boobies also eat many dolphinfish and green halfbeaks. Brown and red-footed boobies consume many juvenile goatfish. Despite such similarities, the diets of Hawaiian boobies differ in several ways. Masked boobies take much larger prey than red-foots or browns, and two or three fish will usually fill a stomach. Browns, by contrast, eat numerous small inshore fishes and their stomachs contain an average of twelve prey items.

Flying squid represent a greater proportion of the diet for red-foots (one-quarter) than for masked or brown boobies (one-twentieth). Red-foots have enlarged eyes that facilitate nocturnal vision and enhance their ability to capture squid. Red-foots often depart the colony to feed well before daylight and are notorious for being the last seabird to settle down at night, especially when the moon is full. Many of their daylight hours are spent in sleep at the colony. The differences in diet among the various booby colonies are fairly minor, but diets change dramatically with season. Pacific sauries, needlefish, rudderfish, and anchovies are important food sources for boobies during fall and winter at Midway and Kure but are seldom eaten during spring and summer.

Boobies on Land

Boobies spend more time ashore than most Hawaiian seabirds and are well adapted to terrestrial as well as marine existence. Red-foots are the only member of the family that can perch, and they are perfectly at home swaying among beach heliotrope limbs during a storm. Nonbreeding red-foots roost at nest sites or on tall ironwood or palm trees. Masked and brown boobies are reasonably mobile on flat open terrain, but browns labor on takeoff, often striking the soil or vegetation with their wings on their first several strokes. Clubs of unemployed boobies may form on breeding and nonbreeding islands at any time of the year. They can include juveniles and adults and usually number ten to thirty birds, but masked boobies on Laysan form clubs of almost two hundred.

Breeding Season and Nest Sites

Hawaiian boobies are residents and are found ashore in colonies during all months, particularly at night. Breeding is fairly synchronous, although less so among brown boobies. Most islands have distinct annual breeding regimes, which may vary a month or so depending on local conditions. All three species breed earliest on Nihoa. Adults sometimes take a rest year between breeding attempts to ease the physiological stresses of reproduction. Individual pairs or small groups may begin breeding in any month, but most boobies breed during spring and summer, red-foots a little earlier than masked and brown boobies. Oceanographic and weather conditions can delay the onset of laying. The nesting cycle at Kure was delayed a month after heavy storms in December 1964 and March and April 1965. Half of the early nesting attempts of masked boobies failed that year.

Masked and brown boobies in Hawaii nest in loose aggregations on low sandy beaches and rocky ground, as they do throughout the world. Neither bird is highly colonial. They require more or less open terrain and probably cannot nest in forests or within tall vegetation. On Kure and Eastern Island, Midway, both species are confined to open areas inland. Male masked boobies clear bare circular patches; Robert C. Murphy called them a poor apology for a nest. On Kure their nests average eight meters from neighbors. On larger islands such as Laysan and Lisianski, masked boobies tend to nest around the perimeter, although some are also found inland. On small, sandy islets such as French Frigate Shoals and Pearl and Hermes Reef, masked boobies usually nest on upper sand beaches, but this preference is not always obvious because the islets are so small.

Brown boobies construct their nests of twigs, grass, driftwood, or debris, often at the inner edge of beach magnolia or under clumps of bunchgrass at the crests of low ridges. On Nihoa, Necker, Gardner Pinnacles, and Kaula they build nests on upper slopes and ridges. Brown boobies can almost be considered to be cliff nesters, apparently preferring sites that overlook sharp drops in elevation, which facilitate takeoff. A substantial nest is important in such a habitat, where an egg could easily roll away from a bare site.

Red-foots in Hawaii nest exclusively on top of shrubs, aided by the evolution of short legs and small, light bodies. They prefer plants a meter or so in height but can nest mere centimeters from the sand or soil if larger plants are unavailable. They do not nest on such islands as Manana, which lack suitable vegetation. The vegetation that supports red-foot nests varies from island to island, but these birds prefer beach magnolia and make good use of it at Midway, Kure, and Lisianski. Where beach magnolia does not grow, virtually any plant will do: *Pluchea* on Laysan; nightshade on Pearl and Hermes Reef; beach heliotrope on Tern Island, French Frigate Shoals; loulou palms, ilima, and ohai on Nihoa; goosefoot on Necker. Red-foots nest on mesquite at Ulupa'u Head and Kilauea Point in the main islands. Their rough nest platforms are about thirty centimeters in diameter and are a meter or two apart. Red-foots typically construct nests of twigs and sticks from beach magnolia and puncture vine, but almost any available plant material may be used. Red-foots and great frigatebirds often nest in proximity because the two species select similar sites.

Courtship and Incubation

Hawaiian boobies begin to spend more time ashore during their third year but most do not breed until they are four. Unlike some gannets, boobies attempt to breed during the first season in which they establish an adequate nest site. Established pairs have some fidelity to each other and to their territory. Pairs often endure for several years, yet "divorce" is so common that only about half of the masked pairs remain together in succeeding seasons. Masked and brown boobies frequently shift territories within a colony. On Kure, 90 percent of masked boobies establish territories that are a substantial distance from the previous year's site.

Males select and defend breeding sites, including the airspace above them, against other boobies. During courtship masked boobies defend at least four times as much territory as they will use after the eggs are laid. Masked and red-footed boobies confine their sexual advertising displays to the ground, but male brown boobies also display in flight. Red-foots defend not only their platforms but also several nearby perches, which they use for landing, takeoff, and roosting. Boobies attract mates by a complex series of exaggerated, stereotyped head and wing movements, including bowing, strutting, and parading. Unmated males advertise to females by sky-pointing—stretching the neck, pointing the bill upward, lifting the wings, and whistling. They repeat such movements several times, and females respond by approaching and emitting drawn-out whistles.

Mated boobies usually return at least three weeks before an egg is laid, the males arriving before the females. The birds recognize one another by sight and voice. Masked and brown boobies parade about together and perform elaborate ceremonies to construct symbolic or functional nests. A male will make hundreds of short excursions beyond his territory to find and return with such token objects as a dirt clod, a feather, a twig, or a pebble in the tip of his bill. He paces back to his mate and places the object in front of her with elaborate and exaggerated care. The female masked booby participates in "nest building" to a lesser extent; she collects only a small fraction of the material.

A red-foot male collects twigs and sticks and brings them to the nest site, where the female constructs or repairs the platform. New material is added throughout the incubation period to replace items stolen by marauding frigatebirds. Established masked and red-foot pairs rub bills and preen each other around the head and the neck. The brown booby male sometimes inserts his bill down the throat of his mate, symbolically feeding her as though she were a chick. Boobies restrict such pair-bonding activities, including mating, to their territories. As among most Hawaiian seabirds, the nest-building activities of neighboring boobies are usually synchronized.

All booby eggs have pale-blue shells with a chalky surface that is readily scratched by adult toes. Masked and brown boobies usually lay two-egg clutches, though they may lay only one egg or as many as three. Brown and masked boobies sometimes roll abandoned eggs from adjacent territories into their nests so that some clutches apparently consist of five eggs. Masked boobies on Laysan occasionally incubate abandoned albatross eggs, glass balls, plastic fishing floats—almost anything. At such times they seem to be well named.

The second egg of masked and brown boobies is laid four or five days after the first. If both hatch, the vast majority of second chicks are found trampled in the nest or dying just outside its perimeter within a week. Sibling murder occurs even when food is abundant and the larger chick is well fed. Thus, in an evolutionary sense, even masked and brown boobies attempt to raise but a single chick. Most second eggs have a chance only if the first does not hatch or if the first hatchling dies immediately. Red-foots invariably lay a single egg; the few instances of apparent two-egg clutches have resulted from the laying of a fresh egg in a nest containing an addled one. All boobies can lay again within three or four weeks if the clutch is lost, but only about half do so. Nevertheless, some brown boobies have layed five times in the same season. When a loss occurs late in a season, the reproductive effort is usually abandoned. Red-foot eggs weigh 58 grams and represent less than one-twentieth of the female's body weight. A masked or brown multi-egg clutch accounts for about 7 or 8 percent of her weight.

Once an egg is laid, red-foots stand their ground tightly when approached, ruffling their feathers, uttering raucous squawks, and lunging toward human eyes. Incubating browns, by contrast, typically take instant flight when alarmed, sometimes kicking eggs from the nest as they launch. Boobies lack

Red-footed boobies

brood patches. When a booby is incubating, it wraps its toes and webs over the eggs and supports its own weight on its outer toes. When an adult settles down, its dense breast plumage covers its feet as well as the eggs. During hot weather, parents shade the eggs on top of their webs to protect the growing embyros from excessive heat. The incubation period for masked and brown boobies averages about forty-three or forty-four days, but the first egg takes about a day longer to hatch than the second. Red-foots incubate for forty-six days. Although both parents incubate, males take somewhat longer bouts. The shifts of masked and brown boobies average ten to thirteen hours, depending on the island, and those of red-foots average twenty-four hours. Masked boobies usually spend only five or six hours feeding at sea, then return to stand like sentinels near their incubating mates. Incoming birds call loudly to announce their arrival but otherwise boobies have minimal nest-relief ceremonies.

Raising the Chick

Hatching success varies widely by island and year. If multi-egg clutches are considered as a single breeding effort, a high percentage of booby eggs hatch: masked, 61–79 percent; red-footed, 68–80 percent; and brown, 73–88 percent. Hatching success is considerably greater in Hawaii than in the Galápagos, Seychelles, and Ascension, where often only one egg in three hatches. A primary cause of egg loss is infertility, but some red-foot eggs are eaten by frigatebirds and others are inadvertently destroyed when frigates steal nesting material. Most of brown boobies' hatching failures stem from desertion or from nesting attempts late in the season.

Newly hatched boobies weigh forty to sixty grams and have naked skin in various shades of purple, pink, and gray. During their first few days, the ugly hatchlings are so weak that they can barely move their heads. Masked and red-foot parents brood their tiny young atop their webs until the chick has grown so large that its head peeks out from under the adult. Eyes open within a few days and feathers grow so quickly that within a week the chick is covered with a sparse, fine white down. The down lengthens and thickens for another three weeks, transforming the chick into a fluff ball that can regulate its own body temperature. Once a chick can be left alone, brooding ceases and the parents can forage simultaneously. Downy red-foots sleep for long periods with their long necks and heads hanging out of the nest, appearing to be quite dead. Nestlings defend their territories against intruders, thrusting with their bills and clamoring loudly. Biologists are wise to approach young red-foots from the windward because of their propensity to defend themselves by voiding showers of filth with deadly accuracy.

Heat stress brought on by the subtropical sun can be a serious problem for boobies. Hatchlings are vulnerable to direct sunlight and may die after only twenty minutes of exposure. Even newly hatched chicks flutter their gular sacs in an attempt to cool themselves by evaporation. Adults and young place their backs to the sun and droop their wings to aid convective heat loss. Once mobile, the young seek any shade near the nest site.

On Laysan boobies are fed every sixteen to eighteen hours. Parents returning to the colony in the evening face ambush by ravenous great frigatebirds. Boobies jettison many meals when frigates overtake them, seize them by the tail, and flip them over. Masked boobies sometimes return in squadrons of three or four, flapping and gliding in perfect unison like B-52s attempting to run a blockade. When parents arrive at the nest site, nestlings crouch, hold their heads back, and call continuously, bobbing their bills up and down and turning their heads from side to side. A chick will jam its entire head into the parent's gape to swallow whole fish and squid directly from the crop. Begging increases in intensity as chicks grow older, and fully feathered chicks can precipitate such a frenzied onslaught that parents may limit feeding visits to under a minute. Sometimes chicks will beg quietly and persistently even when their parents are absent or when neighboring birds are being fed.

Although the time varies with colony and year, primary feathers erupt around the fourth or fifth week (browns), sixth week (red-foots), or sixth to seventh week (masked). Young birds fatten up and slightly surpass adults in weight after about seven weeks. By the time they fledge, chicks have slimmed down to adult weight as a result of the energy demands of feather growth and wing exercise. Once fully feathered, the young practice flying by making extended hops with outstretched wings. Fledging on Kure takes about 91 days for red-foots, 95 days for browns, and 123 days for masked. Masked boobies' down is completely gone within 100 to 105 days, but their wing feathers are not fully grown for another two weeks, at which time sustained flight is possible. Masked boobies require about the same amount of time to fledge the world over. Hawaiian browns and red-foots seem to take a month less to fledge than birds on Christmas (Indian Ocean), Ascension, the Seychelles, and the Galápagos.

All tropical boobies continue to feed their young after they have fledged, up to six months in extreme cases. Hawaiian boobies do so for one or two months. The transition period allows young birds to remain partially dependent on their parents for food before they have to fend for themselves altogether. Immatures spend time at the fringes of the colony or practice diving for prey by day and return to the territory to beg each afternoon. Once they are truly independent, most young brown and masked boobies leave the colony for extended periods of time and do not return until their third or fourth year. Red-foots spend less time at sea and typically return to roost ashore after only a few months' absence. Adults continue to defend the territory for two or three weeks after the young are independent, then usually leave the colony.

Boobies' fledging success in Hawaii is high in comparison with that of colonies on Ascension, the Galápagos, and the Seychelles. Although success varies with island, season, and location within each colony, the vast majority of chicks of each species survive: masked boobies, 79–83 percent; brown boobies, 68–100 percent; red-footed boobies, 84–97 percent. Elsewhere survival of one bird in four is remarkable, although the survival rate of brown boobies on Johnston Atoll is similar to that of their relatives in Hawaii. Overall breeding success (percentage of clutches yielding young) is also high in Hawaii: masked boobies, 57–86 percent; brown boobies, 66–82 percent; red-footed boobies, 66–76 percent. Brown boobies were so successful on Laysan in 1979 that two pairs successfully raised two hatchlings each. Experienced pairs usually nest earlier in the season and have higher success rates than newly formed pairs. Insurance-policy eggs and replacement clutches greatly increase productivity. On Kure, as much as one-fifth of masked and brown young come from the second egg and one-quarter result from replacement nests.

Few predators can take eggs or young from birds as strong and as well armed as boobies, but a few masked boobies lose eggs on French Frigate Shoals when green sea turtles lumber ashore to excavate their own nests. Starvation is directly or indirectly the source of most chick loss. Adult masked and red-footed boobies maintain their body weight even when feeding conditions are so poor that their chicks are starving. As an overall reproductive strategy, adults produce more young in a lifetime by abandoning nesting efforts in poor years instead of putting their own lives at risk. Theft of young by frigatebirds is common, especially when humans chase adults from their nests.

Conservation

Hawaiian boobies have been disturbed by humans at many colonies. Wartime activities on Midway led to the demise of both ground-nesting species. Browns were the most common booby there in the 1930s, but the military requisitioned their nesting areas during World War II. About eight pairs continued to nest on Eastern until 1958, but since the mid-1960s no brown boobies have nested and few have been sighted on Midway. Only a handful of masked boobies nest there today. Red-foots roosted and possibly nested on Sand in the 1920s but today are confined to Eastern.

The use of Kaula as a U.S. Navy gunnery site has killed thousands of boobies. The rocky island is so littered with unexploded ordnance that it is dangerous for biologists to attempt to assess the effects of bombing. Errant shots kill nesting adult and young red-foots in the colony at Ulupa'u Head, Kaneohe Marine Corps Air Station, where birds share a crater with a gunnery range. On a positive note, red-foots established a new colony on Tern Island in the 1970s when shrubs were allowed to grow along the runway. During the 1950s, red-foots at French Frigate Shoals limited their breeding to East and Trig islands, the locations of the only two shrubs on the atoll. Red-foots seem to be in the process of reclaiming some of their former breeding range in the main islands. The colonies at Ulupa'u Head, Kilauea Point, and Sea Life Park were established in 1946, the early 1950s, and the early 1970s, respectively. Undoubtedly red-foots had many additional colonies in the main islands before the arrival of humans. The decline of subsistence hunting and the increase in wildlife protection have laid the foundation for the recent recolonizations.

Introduced predators are a problem for ground-nesting seabirds on tropical islands throughout the world, even birds as large as masked and brown boobies. Cats have eliminated boobies on the main island at Ascension, killing adults and juveniles. Rats prey on booby eggs, young, and adults on Kure and Midway and may have contributed to the demise of the brown booby colony on Eastern. Mongooses attack and kill young red-foots on their nests at Ulupa'u Head, but the shrubs raise the nests well off the ground and afford some protection. Pigs trampled and destroyed masked and brown colonies on Clipperton. Boobies are resilient, however. Within thirty years of the removal of the pigs, the masked and brown booby colonies rebounded to populations of tens of thousands, perhaps again the largest on earth.

Boobies are also affected by changes in vegetation. The absence of red-foots at Manana can probably be attributed to its lack of shrubbery on which to nest. The alien European rabbits died out there in the mid-1980s, and red-foots could reestablish a colony once shrubs return. Masked and brown boobies require open space for nesting. Clearing the antenna field on the central plain of Kure in the 1960s increased nesting habitat for brown and masked boobies. However, human activities on Kure have also brought the introduction of such plants as wild mustard and golden crown-beard, which crowd out native vegetation. Unless such exotic vegetation is controlled, nesting habitat will be reduced and the masked and brown booby colonies on Kure may be lost.

Human consumption of booby adults, eggs, and chicks on the islands and atolls of the western Indian Ocean has caused the populations of all species to decline there during the past century. At least half of the colonies have been eliminated and the remainder are greatly reduced. Hunting is especially difficult to control there because birds are taken for home consumption rather than commercial purposes. Although human consumption has not been an important problem in Hawaii for many decades, it continues to plague boobies in many Pacific island nations, including portions of Micronesia that are under the control of the United States government.

Several fisheries in Hawaii might harm boobies. Commercial-size mackerel scad is an extremely important element in the diet of boobies, especially masked, and the amount caught should be strictly controlled. Fisheries for flying squid or bait fisheries for juvenile goatfishes and anchovies near breeding colonies also should be carefully managed to avoid untoward effects on boobies' food supply.

Boobies are exposed to pollution by heavy metals and chlorinated hydrocarbons in the atmosphere and in the marine food chain. Low levels of DDE and PCBs occur in most red-foot eggs throughout the Hawaiian Islands, an indication that Hawaiian waters have not escaped the general pollution of the oceans during our era. Mercury is also common in red-foot eggs and is highest at Midway. Although the present levels are not grounds for immediate alarm, they are higher than in the eggs of sooty terns and wedge-tailed shearwaters and are similar to levels found in coastal birds in industrialized parts of Europe and North America. If mercury levels were to increase, reproduction might be impaired. The source of the mercury is a puzzle, as there is no industrial or military activity in Hawaii that discharges mercury into the marine environment. Some mercury is introduced into the ocean from deep-sea volcanic activity, and another source may be industrial waste from Japan. Pacific sauries, a favorite food of red-footed boobies at Midway, enter Hawaiian waters during winter, when the Kuroshiro current moves south. Mercury may become concentrated in sauries when they are in Japanese waters and turn up in red-foot eggs when the birds eat contaminated fish at Midway.

14 TROPICBIRDS
Family Phaethontidae

I saw my first red-tailed tropicbirds early one February just after checking into BOQ Alpha on Sand Island, Midway. Stretching my muscles after a long C-141 flight from Honolulu, I strolled along the beach magnolia that fringes the enlisted men's beach on the lagoon. Loud caterwauling drew my attention to a pair of red-tails performing an acrobatic mating dance. The uppermost bird flapped its wings furiously while rapidly switching its long red streamers from side to side, trying to maintain a stationary hovering posture while slowly drifting backward with the tradewind. Meanwhile its mate swooped toward the beach in a long, shallow dive. As the lower bird pulled up out of its glide into a vertical stall, the upper turned forward to glide, and the birds reversed positions. Calling raucously, they alternated the flapping, circling flight with long straight glides. Between late morning and early afternoon at least one hundred tropicbirds soared and glided over Sand. Against the blue of the North Pacific sky and Midway's turquoise lagoon, the svelte creatures were a vignette of celestial, graceful beauty.

Tropicbirds are probably the most primitive pelecaniforms; fossilized tropicbirds at least 60 million years old have been found. These birds have strong, slightly decurved, heavy beaks with serrated edges. Their external nostrils, unlike those of boobies, are fully open. Tropicbirds' wings are short and stout, and in flight these medium-sized seabirds resemble heavy-bodied terns. The sexes cannot be distinguished in the field except just after the females have laid, when their cloacas are distended. Adults (but not immatures) have two enormously elongated central tail feathers that equal the body length and serve no aerodynamic function. The wispy streamers grow and are replaced continuously, so that breeding birds may have one that is fully grown while the other has just emerged. Their small legs and feet are weak and set so far back on their bodies that all shoreside movements are awkward. Tropicbirds can barely walk and cannot stand upright. When they move short distances, they shuffle forward by pushing with both feet and falling forward on their bellies. Each of the world's three tropicbird species has white plumage punctuated with black bars. Adults' feathers are sometimes suffused with peach, rose, or salmon, tints that result in a satiny, iridescent sheen. The purpose of such coloration is unknown. Although males have pink flushes more often than females, suffusion tends to indicate fresh plumage rather than breeding condition.

Tropicbirds' shrill, discordant screams sounded like bosun's whistles to sailors, hence the name bosun birds. The family name and genus of all tropicbirds, *Phaethon*, is derived from the name of the son of Helios, who chased across the sky in his father's sun chariot but lost control and plunged into the sea when Zeus struck him down by a thunderbolt. The Latin *rubricauda* and *lepturus* refer to the red and white tails of the two species that nest in Hawaii.

Known as *koa'e kea* (white-tailed tropicbird) and *koa'e 'ula* (red-tailed tropicbird) to Hawaiians, both Hawaiian species are almost pure white with black eye stripes. Red-tails possess vermilion, narrow-vaned central tail feathers, blue-gray legs, and black feet. Immatures are white with heavy black vermiculations on their backs. The bills of immature birds are gray-black but become coral-red when the birds reach adulthood. Red-tails in Hawaii are somewhat smaller than birds elsewhere in the Pacific, weighing an average 624 grams. White-tailed tropicbirds, sometimes called yellow-billed tropicbirds, are much smaller than red-tails and weigh about 455 grams in Hawaii. Their white plumage contrasts with solid patches of black on the wings and back. White streamers are poor field identification marks except at close range—white-tails are most easily distinguished from red-tails in flight by the black band near the wrist along the leading edge of the wing. Juveniles have dark bills that turn green-yellow as the birds mature. Red-tails can live sixteen years and white-tails probably have similar life spans.

Tropicbirds are aptly named. They rarely stray from tropical or subtropical seas and are found mostly between 30 degrees north and 30 degrees south latitude. White-tails are found in all the world's oceans but red-tails are limited to the Indo-Pacific. Red-tails range from Madagascar and Mauritius east to the Galápagos, north to the Bonins (Ogasawaras) and Hawaiians, and south to Australia and the Kermedecs. Apparently red-tails do not breed in the extreme southwest Pacific. The world population of neither species is known, but the population of red-tails in the central Pacific is estimated to exceed 75,000 birds.

Red-tails breed throughout the Hawaiian Islands from Kure to Lanai; their largest colonies are at Midway, Laysan, and Kure. In the main islands substantial colonies are located at Lehua and Kaula, and scattered nests are found on cliff faces at Niihau (Kawaihoa Point), Kauai (Kilauea Point), and Lanai (Kaholo Pali). The first red-tail nest on Manana Island, offshore Oahu, was found in 1967, and by the mid-1980s at least ten pairs nested there each year. An increased number of sightings and nests of red-tails in the main islands since the 1970s may signify a recolonization of areas that had been disturbed during the many centuries when tropicbird streamers were prized by subsistence Hawaiians. The total Hawaiian red-tail population is about 50,000 birds, half of which do not breed.

Except for the odd pair nesting at Midway, white-tails do not breed in the Northwestern Hawaiian Islands. In the main islands, white-tails nest in cliffs at Mokoli'i (offshore windward Oahu), Kauai (Waimea Canyon, Kilauea Point, Na Pali Coast), Molokai (Pelekunu Valley, Waikolu, and windward sea cliffs), the island of Hawaii (Kilauea Crater, windward coast), Lanai (Kaholo Pali, Maunalei Gulch, Hauola Gulch), and Maui, where at least 500 and perhaps as many as 3,000 pairs breed. As among red-tails, the nonbreeding and breeding populations of white-tails are about equal. On Kauai, they are commonly observed soaring in Waimea Canyon and along the cliffs at Kilauea Point. White-tails in Kilauea Crater, Hawaii, are sometimes overcome by fumes during eruptions and fall into molten lava.

Tropicbirds consume about 2,000 metric tons of marine organisms in Hawaii each year. Because their populations are relatively small, this consumption represents a tiny proportion of the food taken by the seabird community. Tropicbirds have little direct impact on the marine ecosystem.

Tropicbirds at Sea

Tropicbirds are the most pelagic of the pelecaniforms. They can maintain a flapping flight for long periods without rest, alternating soaring glides with fluttering wingstrokes. Although all four toes are connected by a common web, tropicbirds are poor swimmers. They seldom rest on the water—only about one bird in seven observed at sea is flushed from the surface—but they can float buoyantly with their streamers cocked. Christopher Columbus was ignorant of the biology of these birds. On September 17, 1492, at 36 degrees west in the mid-Atlantic, he wrote in his diary that he saw a tropicbird and therefore must be near land because the species "is not accustomed to sleep on the sea." Columbus was nowhere near land.

Tropicbirds reared in Hawaii disperse far and wide. Red-tails banded on Laysan and Kure have been recovered in the open ocean more than 5,000 kilometers to the southeast. They are regularly attracted to ships, usually circling with graceful, steady wingbeats at altitudes of ten to fifty meters. They often follow ships for a few minutes to an hour. Tropicbirds sometimes alight on mastheads and have collided with vessels in the night, seemingly confused by running lights.

When tropicbirds spot a fish, they hover with the head held downward and the bill pointing to the sea, half-fold their wings, and plummet to the water. Like boobies, they have a layer of air sacs around the neck to cushion the impact with the sea. Tropicbirds use their wings to control the dive—steering, spiraling, twisting, turning sharply, changing direction, and plunging to compensate for the movements of their prey. When the wings remain folded on impact, the bird hits the water with a resounding splash and completely submerges. In many instances, the bird extends its wings just before it strikes the water to break its fall and keep it on the surface. Tropicbirds seize fish and squid transversely in the bill and swallow the prey before resuming flight. They are capable of snatching flyingfish on the wing, and their curiosity concerning ships may actually be an interest in the flyingfish that scatter before an oncoming ship's bow.

Tropicbirds rarely fish within sight of land, even when breeding. Red-tails are fairly evenly distributed in the nearshore and pelagic waters adjacent to the Hawaiian colonies from March to November. Their numbers decline from December to February, when some birds apparently leave the area. White-tails, by contrast, are probably year-round residents, and their numbers at sea remain fairly constant throughout the year. Both species feed by plunging, often from considerable heights. Neither is gregarious, and at sea tropicbirds are among the most solitary of seabirds. About nine tropicbird sightings in ten are of a single bird. Tropicbirds rarely feed among flocks of tuna birds. It would be all but impossible for a tropicbird to keep a fish in view in the midst of a swirl of terns and shearwaters, and torpedo diving would probably result in mid-air collisions.

We know much more about the food habits of red-tails than about those of white-tails in Hawaii. Fish constitute over four-fifths of red-tails' consumption; except for a few crustaceans, the remainder is squid. Red-tails take prey of a broad range of sizes, from tiny stomatopods to balloonfish that appear too large to swallow. Red-tails concentrate on fairly large prey, so that their stomachs contain an average of only four prey items. Flyingfish (especially Linne's flyingfish and *Cypselurus* spp.), flying squid, and mackerel scad are their most common prey. They also eat dolphinfish, truncated sunfish (mostly during summer at French Frigate Shoals), and balloonfish. Like boobies and frigatebirds, red-tailed tropicbirds consume many Pacific sauries at Midway and Kure during the winter months, when sauries migrate south into Hawaiian waters. White-tails probably take similar but smaller prey. Elsewhere white-tails eat small fish, squid, and crustaceans such as crabs.

Tropicbirds on Land

Tropicbirds' locomotion ashore is characterized by lurching movements and awkward landings. Nevertheless, their powerful wings permit them to take flight with ease from flat surfaces. Their vocalizations are less than pleasant. White-tails emit harsh, rasping, peevish snarls. Red-tails have vicious tempers and their cacophonous cries can be unbearable to the ear.

Breeding Season and Nest Sites

Both red- and white-tailed tropicbirds begin their nesting cycles in spring, when birds that have attained at least their fourth year initiate breeding. White-tails remain near the main island colonies all year despite an increase in numbers over land during summer and a decline during fall. The first red-tails return from sea in late February. Tropicbirds require shelter from the sun, and adults nesting in exposed situations may be driven away by heat stress. Both species select nest sites that are at least partially concealed from above. On some islands, white- and red-tails compete with each other for nest sites; the larger red-tails most often win. Yet competition is limited in Hawaii. Red-tails are generally confined to coastal sites, whereas white-tails nest inland in the mountains of the main islands, where their lighter, more maneuverable size enables them to nest in cliffs. The coastal cliffs at Crater Hill, Kauai, and the Kaholo Pali, Lanai, support both species. Cavities there are probably divided up by size, with the smaller white-tails using spaces that are too small for the larger red-tails.

White-tail and red-tail nests are usually enclosed by rock and vegetation, respectively. Red-tails tend to nest under beach magnolia, bunchgrass, ironwoods, or beach heliotrope. Generally they choose sites toward the edge of dense cover, because stems in dense thickets impede their labored maneuverability on the ground. Nine out of ten red-tail nests on Kure are located within three meters of the edge of beach magnolia. Red-tails use only a small portion of apparently suitable habitat—in addition to overhead protection they also require peripheral cover to reduce morning and afternoon light.

Tropicbirds are no more gregarious on land than they are at sea. Red-tail nests are at least a half meter apart, and the average distance is three times greater. They defend only the area that they can reach with their bills from the nest. The concentration of red-tail nests in "colonies" is due to a scarcity of suitable sites rather than any desire to be social. On Nihoa, Necker, Lanai, and Kauai, red-tails nest in rock cavities in cliff faces and on ledges beneath overhangs. On Tern and Midway, their sites are under or adjacent to buildings and other structures.

In the main islands, white-tails nest in crevices, recesses, and hollows in the faces of inaccessible cliffs or crater walls. This habitat is somewhat atypical because in many parts of their world range, white-tails prefer trees. The only white-tail nests that have been found in the Northwestern Hawaiians Islands have been on Midway. A few birds have bred in the crotch of an ironwood tree and next to a house on Commander Row. The preference of white-tails for remote nest sites has severely limited our ability to learn many of the details of their natural history.

Courtship and Incubation

Tropicbirds are ambivalent in their attachment to nest sites and mates. Although many pairs return to the identical site and remain together year after year, sites are generally impermanent and pairing is flexible. Red-tails tend to nest near previous sites, but territorial and pair-bonding behavior is minimal. At Kure, red-tails may lay immediately after returning from sea or prospect for a site for several weeks. The most striking behavior of tropicbirds consists of their dramatic aerial acrobatics. Males initiate courtship flights, which function to establish pair bonds, not to maintain them. Pairing red-tails circle, wheel, and glide together over potential nest sites while calling loudly. Birds tending eggs or young do not display, and usually fly directly from their nests to sea and back again. As the season progresses, only nonbreeders display at red-tail colonies. Flurries are especially common in October, just before juveniles migrate.

At the onset of the breeding season, white-tails hover near prospective nest sites along cliff faces, sometimes alighting momentarily before gliding away. Courting mates fly in tandem near rocky promontories, calling back and forth with a series of chuckling sounds. They remain parallel, the male above the female, a wing's length apart. A male may direct his streamers downward to touch his mate. Then they soar together in long, shallow glides while their wings remain horizontal. Groups of five or six white-tails may alternate rapid wingbeats with long glides. Flights usually terminate when one bird enters a nest cavity, where it may be joined by others. Neighboring cavities are often occupied by pairs that lay synchronously.

Red-tails have somewhat more complex displays. Groups of up to twenty birds face into the prevailing tradewind, calling rapidly and harshly to one another. The complete display consists of one to three backward, vertical, interlocking circles at heights of up to one hundred meters. As I observed during my first visit to Midway, alternating vertical circles are repeatedly executed in two parts by displaying red-tails.

During the weeks when tropicbirds are performing their acrobatics, they spend time on the ground. Male red-tails select and prepare the nest site by removing twigs with their bills and kicking out sand or soil. If another male challenges a prime nest site, red-tails may engage in combat for hours. Completed scrapes vary from shallow depressions in the sand to hollowed-out structures that are lined with dry beach magnolia leaves. The birds copulate at the nest site.

All tropicbirds lay single, relatively large eggs. The eggs of white-tails (40–43 grams) and of red-tails (70–72 grams) account for about one-tenth of the female's body weight. Eggs of each species vary in color from pale fawn to purplish black, depending on the amount and arrangement of pigment. Usually an egg has a light-purple ground color with heavier pigmentation at one or both ends. Perhaps because of such variation, red-tails are the only pelecaniform that can recognize and retrieve their own eggs. Females remain at the nest site a full day before laying and incubate only until the male arrives, usually soon thereafter. Tropicbirds, lacking incubation patches, tuck the egg well beneath their abdominal feathers, with their wings slightly spread. Tropicbirds' feet, unlike those of boobies, are too small to play an important role in incubation or brooding. Red-tails have five to eight incubation bouts about a week each. There is little ceremony during changeovers. An incoming red-tail faces its mate and simply crowds it off the egg. White-tails have a brief affectionate ceremony that includes vocalization. Red- and white-tail incubation periods are about 43 days and 41 days. Young birds pip the interior of the egg three or four days before hatching, causing a star fracture to appear and the loss of a great deal of water.

Red-tail (and probably white-tail) chicks begin to hatch in early April. Hatching success is fairly high for red-tails, ranging from about two-thirds to four-fifths. On Kure, most losses of red-tail eggs are attributed to Polynesian rats, which eat the eggs or drive incubating adults away. Occasionally eggs are destroyed during territorial fights between neighbors. Tropicbirds can lay replacement eggs, which are smaller than the initial eggs and require three to twelve weeks to form.

Raising the Chick

Unlike baby boobies and frigatebirds, tropicbird chicks hatch with a coat of long, fine down that varies in color from white to light brown. Parents brood hatchlings beneath their bodies for the first several days, then protect them under an outstretched wing. After two weeks adult attendance becomes increasingly erratic. By the third week the parents return to the nest for the sole purpose of feeding their chick. They fiercely guard their young when humans come near, holding their ground, calling stridently and indignantly. Nestlings cackle and hiss when approached.

Parents begin to feed their young within hours of the hatching and do so an average of every seventeen hours during the chick's growth and development. Arriving red-tails fly directly to the area over their nest, fold their wings, and drop through the vegetation. When a parent moves toward a nestling, the young bird begs with a shrill persistent rattling call and lunges at the adult's bill. The parent regurgitates food almost immediately and usually leaves the nest site within a few minutes. Unlike other pelecaniforms, adults feed their young by thrusting their bills deeply into the throats of the gaping chicks.

At seven weeks, red-tail chicks begin to maintain their nests, removing twigs and other debris that may blow into the scrape. A red-tail chick attains adult weight after about six weeks and may eventually weigh over a fifth more than its parents. It begins to exercise its wings at about eleven weeks and becomes restless, wandering between the beach and the nest site. Some chicks lose track of their parents and starve to death. Hawaiian red-tails fledge at twelve to thirteen weeks. For several days before the chick fledges, the parents' feeding becomes sporadic and fledglings lose weight. Flying young are at or slightly above adult weight. We have little information concerning the phenology of white-tails in Hawaii, but at Ascension they fledge in ten to twelve weeks. Young tropicbirds remain near their natal islands for several weeks after they are capable of independent flight but are no longer fed by adults. Fledglings leave the colonies as early as June, and the last are usually gone by late October.

Survival of tropicbird young varies. On French Frigate Shoals, about three-quarters of the red-tail chicks fledge, including virtually all birds that survive the first month. On rat-infested Kure, by contrast, success can be as low as one-quarter and rarely exceeds one-half. Rat predation is worst on Kure between March and May and essentially stops in June, when the rats' major food source, beach magnolia berries, become abundant. The survival rate at French Frigate Shoals is probably typical of ratfree colonies in Hawaii, where about six eggs in ten yield fledglings. Although good information is lacking, the white-tail rate is probably similar.

Mortality is highest during incubation and when chicks are small. Where rats are absent, storms, heat waves, adandonment, and infertile eggs are the most common causes of loss. On Tern Island, the nesting activities of green sea turtles destroy some chicks. Avian pox, a viral infection that is common among albatross chicks, has infected young red-tails at Midway since 1961. Facial lesions erupt around the eyes during spring and summer and apparently cause some deaths. Although territorial fighting for nest sites kills many chicks at Ascension and the Seychelles, it rarely causes nesting failures in Hawaii.

Adult-juvenile pairs of red-tails have been seen far at sea calling to and answering each other. Although no feeding has yet been observed, such observations suggest that parents offer some form of care to their postfledgling young at sea. Once fledglings depart the colony, they return only after they have acquired adult plumage, two or three years later.

Conservation

The long tail feathers of tropicbirds were probably used by ancient Hawaiians for ornaments, just as they are employed today on some South Pacific islands. Feathers can be pulled from incubating or brooding birds without a great effect on their survival. Red-tails were a favorite among feather poachers in the Northwestern Hawaiian Islands at the turn of the twentieth century, when thousands were lost to the millinery trade. In the Phoenix Islands, red-tails are still used as carrier pigeons, carrying messages attached to their legs.

A prime concern for the management of red-tails is the loss of nesting habitat which might emanate from devegetation or the introduction of predators. Nesting requires adequate shade. When Laysan was desolated by alien rabbits in the 1920s, it became impossible for red-tails to nest there. The rabbits that had been present on Manana until the mid-1980s probably precluded the growth and development of sufficient vegetation to permit red-tails to form a large colony. Red-tails may increase substantially on Manana in the coming years if shrubs proliferate. Rats and other predators eat nesting adults, eggs, and young. Most of the red-tails that bred along the coasts of Kauai until the nineteenth century have been eliminated by predators, except for those that nest in cliffs that are inaccessible to all but the most agile rats. Black and Polynesian rats on Midway and Kure have diminished red-tail populations. Nevertheless, the net result of human intervention in those atolls has been an increase in nest sites. On

Midway, red-tails were more abundant on Eastern than on Sand in 1907. Today the reverse is true—a result of improved nesting habitat after a massive introduction of plants on Sand and of alien rats on Eastern. White-tails once nested in all of Oahu's cliffs, but are rarely sighted there today. David Woodside believes white-tails are declining along the Kaholo Pali, Lanai, for unknown reasons. Their remote nest sites should provide adequate safety from predation, so diseases associated with alien birds are the likely culprits.

Tropicbirds are less likely to be affected by human activities in the ocean than on land. Their offshore and solitary feeding habits make it unlikely that oil or chemical pollution could seriously affect an entire population. Yet tiny plastic fragments are fairly common in red-tailed tropicbirds' stomachs. The plastic particles are so small that they must be contained within prey organisms when they enter tropicbird bodies. Large-scale fisheries for species that account for a substantial portion of the diet, such as flying squid and mackerel scad, could affect red-tails. Despite reports of starvation at tropicbird colonies elsewhere, Hawaiian tropicbirds seem to be well fed. In the absence of major fisheries near the colonies, food supplies do not seem to limit tropicbird populations.

15 TERNS AND NODDIES
Family Laridae (Subfamily Sterninae)

From a sailboat just offshore Waikiki during spring, Kapiolani Park seems to swarm with small white birds fluttering among the treetops. Their ethereal grace and beauty make white terns readily distinguishable from the domestic white pigeons that lumber through the park in search of handouts. Curious and confiding, they seek out and follow people. Flocks of three or four birds will circle around and hover over human heads, bleating, croaking, and chattering in low voices. Considered sacred by indigenous peoples on islands as farflung as the Cooks, the Australs, and Niihau, white terns are the most photogenic of Hawaii's seabirds. If white terns are holy, Midway must be heaven. Thousands of white terns roost and nest in the ironwood trees, on the ledges of buildings, atop telephone boxes.

Terns are slender, graceful birds with long, pointed wings and straight, tapering bills. Their steady wingbeats and buoyant movements have earned them the name sea swallows. Although web-footed, terns swim poorly because their feet are too small and weak for efficient propulsion. Sooty terns spurn the water almost entirely and, for the first five years of their lives, the land as well. Sooties take their food on the wing; they very rarely touch down on the ocean. They live aloft, presumably sleeping while soaring at altitudes where constant attention to flight is not required. Hawaiian terns are fairly drab. The skin on their feet, bills, and legs is usually a uniform black, and their plumage is predominantly black, gray, and white. As among many marine birds, the sexes cannot be distinguished except when a distended cloaca reveals a hand-held bird to be female.

Terns existed in the Lower Eocene, some 60 million years ago. Placed by taxonomists in the order Charadriiformes, family Laridae, subfamily Sterninae, they are closely related to gulls. Authorities differ, but Peter Harrison's compilation of seabirds of the world lists forty-two species of terns and noddies in seven genera. Six species breed in Hawaii, including three terns and three noddies. Noddies are distinguished from terns by their distinctive nodding and bowing behavior during mating ceremonies, but "tern" properly includes noddies.

Sooty terns are strong and feisty for their 198 grams. Their upperparts are mostly jet black, in sharp contrast with their immaculate white underparts. Black stripes pass through the eyes to the base of the bill and separate the white forehead from the throat and side of the head. The tail is very deeply forked, with elongated white-edged outer feathers. The buff-tipped back feathers of fledgling sooties give them a speckled appearance. The Hawaiian name *ewa'ewa*, "to make uncomfortable," refers to the incessant screeching cries made by flocks of sooties at their colonies. Gray-backed terns weigh 146 grams. They are somewhat similar to sooties but have gray backs, wings, and tail feathers. Gray-backs also have much more white on the forehead behind the eye. Juveniles resemble adults except that their backs are grayish brown. Their Hawaiian name, *pakalakala*, probably comes from their favorite food, five-horned cowfish. White terns, also called fairy terns, weigh about 111 grams. Their ivory plumage is broken only by a circlet of black feathers around their dark-blue eyes. Surprisingly, the white feathers cover shiny black skin. Their blue-black bills are slightly upturned when they perch. Juvenile birds often have light-buff markings on an otherwise creamy-white plumage. Hawaiians on Niihau call white terns *manu-o-ku*, or bird of Ku. Paradoxically, Ku is the ancient Hawaiian god of war, one of the four great gods brought to Hawaii from Tahiti. Some linguists suggest that *manu-o-ku* is derived instead from *ohu*, meaning fog, mist, a cloud, or exhaled breath on a cold morning.

Brown or common noddies are dark brown with prominent gray-white crowns and long, wedge-shaped tails. The name of the genus, *Anous*, is Greek for "unmindful," apparently in reference to noddies' tameness when they are approached by humans. Young birds resemble adults except that their white caps are smaller. Brown noddies weigh an average of 205 grams, almost twice as much as black noddies, whose species name, appropriately, is *minutus*. Black noddies are dark with conspicuous white caps, long pointed bills, and slightly forked tails. These birds have two color phases in Hawaii: in the light phase, black noddies are a dark gray-brown with pale-gray tails; in the dark phase they are a uniform sooty black. Unlike black noddies elsewhere, the Hawaiian population has yellow-orange legs and feet. Hawaiians call both black and brown noddies *noio*, but *noio koha*, or "large *noio*," refers only to a brown noddy. Blue-gray noddies, also called Necker Island terns, are the smallest of the world's terns, weighing only 58 grams. These handsome birds are various shades of bluish gray and have slightly forked tails. Blue-gray noddies lack a Hawaiian name, so they may not have appeared in the main islands when humans occupied them.

Hawaiian terns are fairly easy to distinguish in the field. The snow-white plumage of white terns is unmistakable. Blue-gray noddies are unique in their diminutive size and close association with rocky islands. Only brown and black noddies are uniformly dark with white caps. Browns are larger and have longer tails than black noddies, and in good light their names are seen to be well deserved. Sooty terns can be identified from a distance by their flocking behavior as they follow schools of tunas and other predatory fish. At sea, gray-backs are so similar to sooties that they may be mistaken at a distance for the far more common sooties.

Like other tropical seabirds, Hawaiian terns are Methuselahs among birds. Sooty terns and brown noddies, the largest of the Hawaiian species, live at least thirty-two and twenty-five years, respectively. White terns, black noddies, and gray-backed terns can live sixteen to eighteen years. The tiny blue-gray noddies live at least eleven years. Blue-grays probably have greater longevity, but their colonies are so inaccessible that few of the birds have been banded or recovered.

Distribution and Abundance

Sooty terns, white terns, brown noddies, and black noddies are widespread in all tropical oceans. They breed or once bred in virtually every island group in the tropical Pacific. Gray-backed terns and blue-gray noddies are confined to the tropical Pacific. Gray-backs are most common in the central Pacific, where they breed in the Phoenix, Line, Tuamotu, Fiji, and Wake islands. Blue-grays breed at most island groups in the central and southern Pacific Ocean. Below 25 degrees south latitude, blue-grays occur in a dark morph and are called blue-gray fairy ternlets.

The population of terns and noddies in Hawaii includes a substantial number of nonbreeding birds, ranging from about half of the brown noddies, blue-gray noddies, and gray-backed terns to almost two-thirds of the sooty terns, white terns, and black noddies. The total population of terns and noddies in Hawaii is almost 9 million birds, and accounts for well over half of Hawaii's seabirds.

Except for blue-gray noddies, each species breeds throughout the Northwestern Hawaiian Islands and Kaula. With a total population estimated at almost 8 million, sooties are the most numerous terns in Hawaii and probably the most abundant species in the tropical Pacific. Over half of Hawaii's sooties nest at Laysan and Lisianski. Brown noddies number half a million and have a somewhat southern distribution, with major colonies at Nihoa, Kaula, and Manana. Gray-backs have large colonies at Nihoa, Laysan, and Lisianski and total about 200,000 birds. Black noddies and white terns total about 90,000 and 80,000 birds, respectively. Midway is the largest colony for both species, and major white tern colonies are also located at Laysan and Nihoa. Because blue-gray noddies require cliffs or rocky outcrops to nest in Hawaii, most of the 16,000 birds are confined to Nihoa, Necker, La Perouse (French Frigate Shoals), and Gardner Pinnacles. A small colony may still exist in the steep cliffs at North Horn, Kaula, but breeding has not been confirmed there for fifty years.

Most terns and noddies have colonies on the main islands. Manana and Moku Manu have substantial populations of sooty terns and brown noddies. Gray-backs breed at Moku Manu and were reported nesting on Niihau in the 1880s but have not been found there since. A few black and brown noddies breed on Mokolea Rock, Kailua Bay. Black noddies nest in *noio* caves or on rocky cliffs on Hawaii (Laupahoehoe Park, Paokalani, Kalapana coast), Kauai (at least ten sites along the Na Pali coast, including Hanakapi'ai), Oahu (Moku Manu, Manana [northeast cliffs], Kaohikaipu), Molokai ('Ilio Point, Kahinaakalani, and windward islets), Maui (windward islets), Kahoolawe, Lanai (Kaholo Pali, Kaunolu Bay), and Lehua. Although Sandford B. Dole included white terns in his 1869 list of the birds of the main islands, Niihau was probably the sole nest site during historical times. Since the 1940s white terns have been seen often at sea near Oahu, especially near Moku Manu. A nest was discovered at Koko Head in 1961 and the Oahu population has since grown to at least one hundred birds, concentrated in Kapiolani Park and portions of urban Honolulu.

Sooty and gray-backed terns migrate from Hawaiian colonies when the breeding season is completed. After the young have fledged in August, sooties spend less time ashore yet remain near the colonies for about a month before departure. Individual sooties regularly move among the colonies in Hawaii and Johnston Atoll, while much of the population apparently shifts to the western Pacific during fall and winter. Many sooties banded in Hawaii have been recovered at Wake, Japan, Christmas, the Phoenix Islands, Fiji, Papua New Guinea, Kwajalein, and the Philippine Sea. White terns and brown noddies seem to be semimigratory. A few white terns roost year round, especially on Oahu, and some brown noddies gather in clubs on beaches or nearshore rocks to doze and preen. The vast majority of adults, however, winter at sea in Hawaiian waters, probably within eighty kilometers of an island. The propensity to remain year round near their colonies combined with their dependence on tuna schools to feed may explain the southern distribution of large brown noddy colonies in Hawaii. There are more tuna schools between Nihoa and Manana during winter than between Laysan and Kure, so that the paucity of winter foraging opportunities may limit the size of colonies at the northwestern end of the chain. Blue-gray and black noddies are residents that roost ashore all year. Nevertheless, interisland movements of black noddies among the Hawaiian colonies imply that some individuals venture far to sea.

Terns and Noddies at Sea

When fishing, terns and noddies constantly scan the surface of the sea. After encountering a shoal of prey, they hover with their bills pointed downward and at a favorable moment close their wings to plunge swiftly headfirst. The six Hawaiian terns have variable oceangoing propensities. Sooty terns are among the most pelagic of seabirds, foraging hundreds of kilometers from their colonies

Sooty terns

when nesting and migrating half an ocean away during winter. In sharp contrast, blue-gray noddies rarely stray beyond sight of their rocky island homes.

Sooty terns' strong, continuous wingbeats can achieve velocities of 45 kilometers per hour. Similar to swallows and swifts in having low wingloading and high aspect ratios, sooties use surprisingly small amounts of energy in flight. They lack substantial oil glands and can become sadly bedraggled and even drown if they alight on water for more than a few minutes. When fishing, brown noddies never soar at high altitudes. They move with forceful, steady strokes three or so meters above the water, hovering to pick up small fry and occasionally executing shallow belly flops. Sooty terns and brown noddies will alight on flotsam: ships' rigging, the backs of turtles, glass fish balls, the head of a swimming pelican. Unlike sooties, brown and black noddies sometimes raft on the water. Black noddies have a more fluttery flight than browns and their wingbeats are faster. White terns seem to fly erratically, flapping slowly and effortlessly above the waves. Blue-gray noddy flocks flutter at the sea surface like butterflies.

Sooty terns and brown noddies typically feed offshore over schools of skipjack tunas, yellowfin tunas, dolphinfish, or porpoises. Large predatory fish drive smaller prey to the surface, where mixed flocks of sooties, brown noddies, wedge-tailed shearwaters, white terns, and great frigatebirds feed. Although sooty terns sometimes feed on moonlit nights elsewhere, there are no reliable reports of such behavior in Hawaii. Brown noddies eat about two-thirds fish and one-third flying squid, while the sooties eat slightly more flying squid than fish. Sooties and brown noddies feed on prey of the same size. The most common fishes in their diets are juvenile goatfish, mackerel scad, and flyingfish, although the latter two are unavailable during winter. Brown noddies also consume many larval Foster's lizardfish near the shore.

White terns commonly feed with predatory fish but breeding birds are restricted to schools that forage close to colonies. Nonbreeding birds sometimes range far offshore. Almost 90 percent of the white terns' diet consists of fish, especially juvenile forms of goatfish, flyingfish, dolphinfish, halfbeaks, and needlefish. Their prey is similar in size to that eaten by sooty terns and brown noddies. White terns seem to consume anything they can land, including at least thirty-three families of fish. Depending on local availability, prey ranges from tiny, threadlike snake mackerel larvae to needlefish the length of a white tern. Prey varies dramatically in composition among colonies and seasons. At Laysan, migratory flyingfish and dolphinfish are not taken during winter but are prominent in the diet during spring. During spring at Midway, where water temperatures remain cool, white terns consume anchovies. During summer, white terns feast on goatfish at Laysan but cannot find any at Midway. Because parents return to nests with prey held crosswise in their bills, fish are often in excellent shape and provide fishery biologists with samples of juvenile forms that are difficult to acquire. Describing Gregory's fish in *Fishes of Hawaii*, Spencer W. Tinker wrote:

> This little fish is known from but a single specimen about two inches in length from Laysan Island which was 'brought to the nest of a white tern' on May 12, 1923. This is an example of the extreme depravity to which scientists will descend to obtain a new species, namely, taking food from a little bird.

Black noddies feed inshore over schools of nearshore predatory fishes such as little tunas and jacks, often within a few meters of the shoreline. They can be seen foraging in the brackish waters of Kaluapuhi (Nu'upia) Pond, Kaneohe Marine Corps Air Station, and Kanaha Pond, Maui. Black noddies eat virtually all fish. Although their prey is about the same size as that of white terns, black noddies also take large numbers of smaller prey. They concentrate on larval and juvenile forms of goatfish, Foster's lizardfish, round herrings, flyingfish, and gobies. Black noddies are particularly opportunistic and change diet with the season and among colonies. Goatfish are rarely eaten at Midway despite their dominance in the diet at Laysan year round. Round herrings are commonly consumed during spring and summer but are not eaten in winter. During winter, black noddies at Midway and Laysan eat bristlemouths, while birds at French Frigate Shoals eat gobies.

Gray-backed terns and blue-gray noddies feed close to their colonies without the assistance of fish schools. When gray-backs leave the colonies in fall and winter, they probably feed far offshore. Ninety-two percent of the gray-backs' diet consists of fish, with flying squid and crustaceans accounting for the remainder. The gray-backs' diet is dominated by juvenile five-horned cowfish, reef fish that are roughly ellipsoid. Such a specialized prey implies the existence of unique feeding areas. Gray-backed terns also eat substantial numbers of juvenile flyingfish, goatfish, round herrings, and dolphinfish, the latter only during spring and summer at the northern end of the archipelago. Striped hawkfish, flying gurnards, bristlemouths, and sea striders are substantial components of the diet at some colonies.

Blue-gray noddies feed individually or in small flocks and tend to be active at first light, returning to roost in the afternoon. They repeatedly dip or patter at the surface without alighting or wetting their feet. Their diet is unique. By volume, about 60 percent consists of fish, with the remainder divided equally between minute crustaceans and microscopic sea striders, a marine insect. Sea striders are found in eight of ten stomach samples and are especially prevalent during spring. Blue-gray noddies' prey are so tiny that their stomachs can contain hundreds of items. Other common foods are stomatopods, copepods (*Pontella atlantica*), and juvenile forms of Forster's lizardfish, flyingfish, and goatfish.

Terns and noddies account for over one-fifth of the marine resources that are consumed by seabirds in Hawaii. Because of their vast population, sooty terns account for most of the food eaten by this family.

Terns and Noddies on Land

Most Hawaiian terns first breed between three and four years of age, but sooties defer reproduction for six or eight years. At an expanding colony on Oahu, some white terns may breed as young as two. On land, sooty terns are notorious for the tremendous clamor made by their large screaming and snarling multitudes. The harsh, discordant ruckus continues at the same deafening decibel level night and day, hence the name wideawake. The great swarms of sooties hovering over Laysan and Lisianski form veritable clouds over the islands during late spring. Brown noddies, black noddies, and gray-backed terns also utter sharp, rasping cries when intruders approach too close to their nest sites but rarely form dense masses of angry birds. White terns and blue-gray noddies make no obnoxious noises.

Breeding Season and Nest Sites

Unlike many populations of tropical terns elsewhere, Hawaiian terns generally have annual breeding cycles. Individual birds may nest in any month and the onset of the season may vary from year to year with oceanographic conditions and the birds' physiological readiness to breed. Nevertheless, certain patterns are apparent over time. For four species—sooty terns, gray-backed terns, white terns, and brown noddies—breeding peaks during spring and summer. Although a few aberrant sooties have laid during fall on Moku Manu, the vast majority in Hawaii lay between March and July. Christmas counts by the Audubon Society in the 1960s have estimated as many as 6,000 sooties on Moku Manu, but they were hovering and circling well offshore—characteristic courtship behavior a month or two before laying.

Gray-backed terns arrive and lay shortly before sooty terns, with whom they compete for nest sites. Gray-backs' egg-laying usually peaks between March and late April, although eggs may be found as early as February and as late as July. White terns have a distinct egg-laying peak in April and May in the Northwestern Hawaiian Islands, a month later than on urban Oahu. Dorothy Miles has found that about a quarter of the pairs in a new and growing colony on Oahu seem to breed at intervals of less than twelve months, so that multiple offspring may be raised in a year. Brown noddies have a protracted nesting season with fairly distinct peaks of egg-laying in both spring and summer. Consequently brown noddy eggs are found from March to August.

Blue-gray and black noddies have winter–spring breeding seasons, which occasionally extend into summer during years of inclement weather. Blue-grays lay their eggs between December and mid-March, and all young fledge by mid-May. Though black noddies lay as early as November, their egg-laying peaks in December and January and continues through June. High winds often destroy winter nests, so that spring laying may actually represent renesting efforts.

Hawaiian terns use many types of nest sites. Males of ground-nesting species probably select a different site each year. Gray-backed and sooty terns lay in small depressions on bare ground, such as sand or gravel. Sooties prefer to deposit their eggs near such vegetation as bunchgrass, which provides shade during part of the day. Gray-backs choose sites with somewhat more cover, such as beneath beach magnolia, bunchgrass, or beach morning-glory. Because gray-backs may lose prime nest sites to sooties, they are often at the fringes of a sooty colony. On Laysan, sooties have crowded gray-backs from the center of the island toward the sea, so that gray-backs form a scattered community that encircles the island along its beaches. On Nihoa and Necker, gray-backs are more prone than sooties to nest on rock ledges and ridges.

Hawaiian brown noddies also nest on the ground, often on open slopes or under such vegetation as bunchgrass or beach magnolia. Although brown noddies typically nest on cliffs in the Caribbean, in Hawaii they do so only on Nihoa. On Sand Island, Midway, some browns nest in ironwood trees as high as twenty meters from the ground.

Hawaiian blue-gray noddies nest only in cliffs and rocky outcrops, preferring sites in the lee of northern storms. Blue-grays avoid isolated cavities and form loose nesting aggregations dictated in part by the presence of clustered cavities within ancient lava flows. Nest holes protect adults and young from direct sunlight. Accumulations of guano at their sites imply repeated occupation for nesting and roosting. Black noddies nest on available cliff ledges and rocky pinnacles throughout the archipelago. Many colonies on the main islands are located in rocky sea caves, where nests are constructed above the surge of high water. Black noddies also nest in such vegetation as ironwood trees, beach heliotrope, and bunchgrass. White terns do not build nests but lay eggs wherever they find a stable surface. Male and female white terns participate equally in site selection by walking back and forth over the branch or ledge, as if to test the substrate for stability. On Nihoa and Necker, the female lays her egg on a sheer cliff or rocky outcrop. Elsewhere eggs are deposited on bare rock or plants ranging from tall ironwood trees to stunted beach magnolia plants mere centimeters above the ground. On Midway and Tern, eggs are laid on virtually any human structure.

Courtship and Incubation

Hawaiian terns are social. Established pairs such as white terns tend to remain together for several seasons and elaborately preen each other about the head. Most ritualized displays, however, are performed by birds forming bonds or mating for the first time. Although all Hawaiian terns seem to be colonial, white terns and brown noddies are found in such low densities that they may be drawn together because of limitations of nesting habitat rather than any desire to nest in groups.

At the onset of the breeding season, sooty terns congregate night after night in dense swirls a few kilometers offshore. In the following weeks they gradually move closer to the colony until they hover above it, often alighting briefly before returning to sea. Such activities synchronize breeding for the entire flock. Once terns settle on the ground, they begin to display. All ground-nesting terns parade by prancing about rapidly with their heads extended well forward and their wings held away from the body. Birds typically parade immediately after landing and just before copulation. Courting pairs engage in "high flights" during which they ascend so high that they become specks in the blue, then glide rapidly to sea level. Males defend their territories aggressively. Even an otherwise docile white tern will attack or threaten any bird that lands within a meter of its site, jabbing with a sharp beak that could inflict mortal injuries.

All terns and noddies have ceremonial fish flights in which one bird presents and sometimes transfers a fish to its partner. The nodding and bowing typical of noddies occurs when a male feeds his mate before she lays her egg. John B. Watson's description of brown noddies' courtship feeding at the turn of the twentieth century remains vivid:

> The male returns with a fully-laden crop. He alights directly upon the nest or near the female. The female at once shows signs of life, and as they

> approach each other they begin nodding. Then the male invites the female to feed by putting his beak down to a position convenient to her. She gets the food by taking it directly from the mouth of the male, the male disgorging it by successive muscular contractions of the throat and abdomen. The impression one gets from this ludicrous performance is that the bird is choking to death.

Few Hawaiian terns build true nests. Blue-gray noddies merely adorn their bowllike cavities with some stray quills, bird bones, or twigs. Sooty and gray-backed terns decorate the perimeters of their shallow scrapes with a few leaves, shells, or pebbles. White terns do not construct even the semblance of a nest. Black noddies, by contrast, build genuine nests, often using old sites for foundations. While the female defends the platform, the male collects twigs, ironwood needles, algae, or feathers, which he compacts with guano.

The female lays a single egg, which represents at least a fifth and as much as 27 percent (blue-gray noddies) of her body weight. Both sexes develop brood patches to incubate the egg. The length of incubation shifts varies widely with location: sooty terns average five days on Laysan but only four on Manana. Gray-backs have eighteen-hour shifts on Laysan but only seven on Tern Island, where a large protected lagoon allows sheltered inshore feeding. Noddy and white tern shifts average about a half and a full day, respectively. Terns in Hawaii have nesting periods between 34 and 36 days except for sooties (29 days) and gray-backs (30 days).

Heat stress is a problem in spring and summer. During intense sunlight, sooties stand astride their eggs to shade them. On Eastern Island, Midway, sooties that nest in the shade of exotic ironwood trees have higher reproductive success because temperatures are cooler by a full ten degrees centigrade. Although most birds incubate their egg continuously, sooty terns and brown noddies sometimes abandon it for a few minutes to wet their breast feathers or to drink seawater by skimming the sea surface with open beaks. Terns have little capacity for evaporative cooling and must rely instead on convective methods. Ground-nesting birds prefer nest sites that are close to shade plants or exposed to cool sea breezes.

Raising the Chick

The hatching success rates of most Hawaiian terns average at least 70 percent, depending on local conditions. White terns are an exception: only a third to one-half of their chicks hatch, usually because of the precariousness of their nest sites. Birds that lay early in the season and in the center of a colony usually fare better than late nesters that have been relegated to the fringes. Most egg losses are caused by infertility, rats, adandonment, or poor site selection. Weather can also be a factor—an April storm in 1970 destroyed over 4,000 sooty tern eggs on Manana. All Hawaiian species are capable of laying again within a few weeks after the loss of an egg and can do so repeatedly. On Ascension, virtually all black noddies and about half of the white terns that lose eggs early in the season lay again, but fewer do so as the season progresses. An egg develops a star fracture three or four days before the chick hatches, and the chick can be heard peeping within the shell.

Tiny nestlings are brooded for at least a few days by black noddies and for several weeks by blue-gray noddies. White tern hatchlings emerge from the shell with well-developed feet and long, sharp claws that are immediately capable of clinging to rocks and bare branches. Ground-nesting parents shade their young during the hot hours of the day and brood only when it is cool. Sooty and gray-backed chicks form crèches when they become mobile and seek shade beneath nearby vegetation.

After about three weeks adults forage at sea and spend little time with their young. Chicks are fed at least twice a day. Upon returning to the nest, the parents are besieged and enticed to regurgitate within minutes. On Laysan, the variation in feeding frequency among most species is surprisingly small, ranging from eleven hours (black noddies) to sixteen hours (sooty terns) between feeds. White tern chicks are fed much more frequently—every three hours—and parents continue to return with food well after dark. Unlike any other tern, white terns carry fish and squid crosswise in their bills and deftly move their tongues to feed as many as sixteen organisms to their young, one by one. The regurgitate from sooty terns is often covered with mucus, which may function to delay digestion of prey obtained at great distances from the colony.

Sooty parents learn the voices of their hatchlings after four or five days; after that they peck at strange chicks. Recognition is important in vast sooty colonies. Parents returning with food visually locate their own scrapes and upon landing may be approached from all sides by begging chicks that emerge from the shade of the bushes. Adult and young white terns, by contrast, do not vocalize to each other.

Individual terns and noddies grow and develop over variable periods of time. Black noddies grow the fastest and average thirty-eight days from hatching to first flight. Sooty terns are laggards, requiring eight weeks to fledge. During intensive growth, young sooties remain close to their nests or scrapes and move only to avoid the ravages of the sun. Brown noddies put on weight quickly—some nestlings weigh as much as adults after only eighteen days, and most are heavier than their parents after six weeks. They are capable of short practice flights before their wings are fully grown and will flee when alarmed, later returning to the nest. White terns begin to take short flights at seven weeks, long before they have lost all of their down, attained their adult weight, or have fully grown wings. Parent white terns may literally push a chick off its perch for the first flight and then fly alongside the fledgling. Within two weeks the fledgling can fish on its own.

All offspring continue to be fed by their parents for several weeks after they begin to fly, but brown noddies remain dependent for over three months. During this transition period, juveniles learn to fish by day and return to the colony in the afternoon to receive a food allowance. Young gray-backed terns remain at the colony for up to six weeks after they can fly, then depart to the sea. Most migratory terns probably do not feed their young at sea, yet pairs of adult and young sooty terns are seen together far offshore during October, incessantly calling back and forth to each other. Parents are seen to feed their young in the air above colonies in the Seychelles. Patrick J. Gould has observed sooty parents sitting on the water with juveniles and feeding them in the central Pacific.

The survival rates of Hawaiian tern and noddy chicks are fairly high in comparison with those of colonies elsewhere in the tropics, averaging at least three-quarters for each species and approaching 90 percent for black noddies, gray-backed terns, and blue-gray noddies. A nest's location affects the survival of the chick. Brown noddy chicks raised in bushes fare much better than those on open ground. Overall survival from egg to young depends on the success of both egg and chick. The fraction of eggs that results in young varies from about a third for white terns to three-quarters for gray-backed terns. About half of the nesting attempts of sooty terns, blue-gray noddies, black noddies, and brown noddies result in young. Extensive egg loss is responsible for the poor success rate of white terns in the Northwestern Hawaiian Islands. On Oahu, where egg loss is minimal, overall success is 80 percent.

Success rates vary widely. Survival of sooty terns on Manana during the 1970s varied between 3 and 84 percent, depending on month and location within the colony. At Tern, virtually all brown noddy chicks that hatch after mid-May may die even when the survival of earlier hatchlings is normal. High winds are a chronic cause of white tern and black noddy losses, and storm-driven high waves wipe out gray-backed tern nests near the surf zone. Starvation affects all species, but massive losses are suspected only for sooty terns (Midway, 1907) and brown noddies (Manana, 1947).

Adult sooties defend their territories by savagely pecking and killing any sooty or gray-back chick that wanders too close. Great frigatebirds, especially females and subadults, eat many small and medium-sized sooty tern, brown noddy, and gray-backed tern chicks. Predation ceases when chicks become too large for frigates to swallow. On Ascension, frigatebirds can recognize brooding sooty terns and will swoop down to grab an adult by the head, cast it in the air, and return to pick up the exposed chick. Frigatebirds cannot maneuver to reach the tree and cliff nest sites of white terns, black noddies, or blue-gray noddies.

Black, Norwegian, and Polynesian rats are important predators of young terns on Midway, Kure, and the main islands. Black-crowned night herons and occasionally cattle egrets eat ground-nesting terns on Manana and Moku Manu but probably do not greatly affect production. Alien common mynahs, bulbuls, and rock doves harass white terns on Oahu, causing some losses. On Midway, mynahs evict black noddies and white terns from their nests and may eat their eggs.

Conservation

Humans have devastated tern colonies throughout the world. Tern eggs are quite edible, and sooties are probably the world's largest producers of wild eggs. The sooty tern colonies at Dry Tortugas, Florida, were almost eliminated during the nineteenth century, when eggs were sold to wholesale bakers. Five million sooty tern eggs are still harvested each year in the Seychelles and Amirantes. Ancient Hawaiians no doubt destroyed most main island colonies long before historical times. To their own detriment, many species are tame. Hawaiians regularly raided black noddy colonies for eggs and young, secure in the knowledge that they could easily approach the unwary birds. Black noddy colonies in the main islands today are restricted to relatively inaccessible cliffs and caves. Gray-backed terns nested on Niihau and Kauai in the 1880s but no longer do so, possibly because humans so often collected their eggs. Feather hunting has also destroyed many tern colonies. Ancient Hawaiians took feathers from many birds, and there is no reason to believe they avoided terns. The historical absence of nesting white terns in the main islands before 1961 may be attributable to feather hunters. White and gray-backed tern feathers were highly prized in the Northwestern Hawaiian Islands during the millinery era, when populations at Laysan and Lisianski declined severely.

Military activities have often posed conservation problems. No blue-gray noddies have been found nesting at Kaula since the U.S. Navy began to use the island as a bomb target in 1952. U.S. warplanes shelled and strafed Manana, Kapapa, and Molokini during World War II, no doubt shortening the lives of nesting birds considerably. Some tern colonies are especially susceptible to disturbance by biologists. Incubating gray-backs on Nihoa and Laysan flee when humans approach, leaving their eggs prey to endangered finches.

Any large tern colony is a potential feeding ground for alien and natural predators. Without islands, cliffs, and remote beaches that are free from predators, terns might not survive. Rats and cats are a scourge for small ground-nesting birds throughout the world and have brought the demise of many tern colonies. Islands with rats tend to have the smallest tern colonies in Hawaii. The populations of brown noddies, sooty terns, and gray-backed terns have declined at Midway since the introduction of black rats in 1943. Gray-backs, smaller and weaker than sooties, seek out the ratfree sandspits between Sand and Eastern islands to nest. On Sand, some brown noddies escape predation by nesting in ironwood trees rather than on the ground, an unknown nesting habitat for them in Hawaii before the arrival of rats. When Polynesian rat populations on Kure are high, the hatching success of gray-backed terns drops remarkably. If rats, cats, or other mammalian predators were ever introduced to Laysan, Lisianski, Nihoa, or Necker, the tern colonies there could be devastated. Even alien insects can create problems. Exotic ants on Midway climb trees where white terns nest and attack the webbing, eyes, and mouths of young birds.

Pollution by heavy metals and chlorinated hydrocarbons does not pose an immediate threat to Hawaiian terns. Sooty tern eggs on Oahu, French Frigate Shoals, Laysan, and Midway lack measurable levels of cadmium and PCBs. Sooty eggs do contain mercury, but less than the amounts found in sooties at the Dry Tortugas and below levels that could cause reproductive failures. Mercury probably enters the marine environment through cyclic volcanic eruptions rather than human activity and it concentrates in feathers, livers, kidneys, and eggs. Low levels of DDE are found in sooty terns.

Certain commercial fisheries proposed for the Northwestern Hawaiian Islands could create food shortages for terns. If the tunas and dolphinfish that forage in the vicinity of the colonies were overfished, the birds that rely on predatory fish schools to drive prey to the surface could have difficulty feeding their young. Bait fisheries for round herrings, anchovies, juvenile goatfish, and juvenile mackerel scad would compete directly with birds. A fishery for flying squid might affect terns, yet the availability of squid is probably correlated more with the presence of predatory fish than with the amount of squid in the water column. Blue-gray noddies would probably be unaffected by any fishery because they eat minute prey and do not forage with tunas.

Terns and noddies have benefited from human contact. When Midway was transformed from a sand dune to an ironwood forest at the turn of the twentieth century, a nesting habitat for black noddies and white terns was created. Black noddy and white tern populations soared from a handful of birds to several thousands. White terns moved from La Perouse Pinnacle to Tern Island in the 1950s once beach magnolia, beach heliotrope, and ironwoods were sufficiently established to provide suitable nesting habitat. Although hundreds of black noddies were found regularly at Kure during spring, summer, and fall, they did not nest there until the 1980s, after beach heliotrope was introduced. The clearance of Kure's central plain to erect a LORAN tower in 1963 allowed an increase in the brown noddy population and colonization by sooty terns from Midway. Though some sooties die each year in collisions with LORAN guy wires, the net result of interaction between humans and sooty terns at Kure has been positive.

Hawaiian terns are resilient enough to adapt successfully to changing conditions. Sooty terns moved en masse from Moku Manu to Manana in 1947 and from Sand to Eastern, Midway, after World War II. At French Frigate Shoals, sooty colonies have moved back and forth between Tern and East several times since the early 1940s. Black noddies and white terns on Midway moved from Eastern to Sand after World War II. The vast brown noddy and sooty tern colony on Manana did not exist in 1900, when Hawaiians were eating eggs and young birds. Once Manana was protected as a wildlife sanctuary, it flourished into one of the largest brown noddy colonies in Hawaii. The ability of terns to change their nesting locations and reestablish their colonies after disturbances have abated may be their greatest asset.

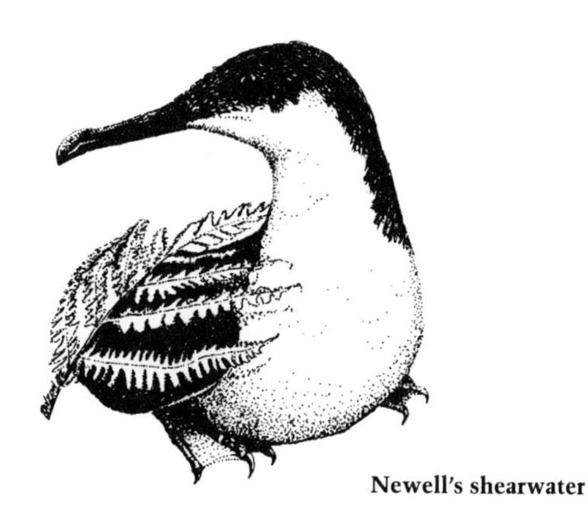

Newell's shearwater

PART IV

Conservation

It is hereby ordered that the following islets and reefs, namely: Cure Island, Pearl and Hermes Reef, Lysianski or Pell Island, Laysan Island, Mary Reef, Dowsetts Reef, Gardiner Island, Two Brothers Reef, French Frigate Shoal, Necker Island, Frost Shoal and Bird Island, situated in the Pacific Ocean at and near the extreme western extension of the Hawaiian archipelago between latitudes twenty-three degrees and twenty-nine degrees north, and longitudes one hundred and sixty degrees and one hundred and eighty degrees west from Greenwich, and located within the area segregated by the broken lines shown upon the diagram hereto attached and made a part of this order, are hereby reserved and set apart, subject to valid existing rights, for use of the Department of Agriculture as a preserve and breeding ground for native birds. It is unlawful for any person to hunt, trap, capture, wilfully disturb, or kill any bird of any kind whatever, or take the eggs of such birds within the limits of this reservation except under such rules and regulations as may be prescribed from time to time by the Secretary of Agriculture. Warning is expressly given to all persons not to commit any of the acts herein enumerated and which are prohibited by law.

This reservation to be known as the Hawaiian Islands Reservation.

—THEODORE ROOSEVELT, 1909
Executive Order no. 1019

16 CONSERVATION ON THE ISLANDS

It was my first overnight on Eastern Island, Midway. I would have preferred a bed back on Sand, but it was not safe to navigate a small boat alone on the lagoon after collecting food samples late into the night. The starlit North Pacific sky seemed friendly enough, but I decided to camp in an abandoned bunkhouse to avoid sudden squalls. Well ensconced on a cot after a day of furious physical activity and a few bedtime swallows of zinfandel, I entered a deep and dreamless sleep. Near midnight I awoke with a start. Vaguely aware that a mass of squirming matter had landed between my legs and was surveying my lower body, I reached for my flashlight. The interloper scurried to the floor to join dozens of its rodent brethren. For the rest of that night and on all subsequent trips I slept exposed on the pier and took my chances with the rain. One experience that harrows the soul is enough. But where can a nesting petrel go? Many important Hawaiian seabird colonies have been designated as parks, refuges, natural area reserves, or sanctuaries. Federal, state, and county statutes protect wildlife from human activities with varying degrees of success, but what protects a small bird in its burrow when the hungry rat arrives?

Many of the most blatant forms of abuse to seabirds have stopped. Young albatrosses on Laysan are no longer dipped in boiling water to permit their down to be stripped off. Sooty tern eggs are not collected by the million for sale in the markets of Honolulu, as they are in the Seychelles, nor is subsistence egging a problem, as it is in the Caribbean. Fledgling shearwaters are not canned for sale as mutton birds or processed for pharmaceuticals, as they are in Australia and New Zealand. Modern threats to Hawaiian seabirds tend to be insidious, and protection efforts must respond accordingly.

Threats

Terrestrial threats to seabirds include three broad and overlapping categories: alteration of habitat, disturbance by humans, and the introduction of alien predators, insects, and vegetation.

Alteration of Habitat

The alteration of a species' habitat often eliminates appropriate sites for its nests. Most seabirds colonize only specific locations, and although such species as terns may relocate, biologists do not know how to encourage them to do so. Large portions of Midway, Kure, and French Frigate Shoals have been altered by construction activities. Although the enlargement of Tern and Sand islands created nesting habitats for some species, the vast surface area occupied by roads, housing, and runways removed many nest sites. Midway was proposed as a site for the disposal of nuclear wastes in the 1970s, which no doubt would bring renewed activity and disturbance to an important colony in addition to the more obvious millennia-long problems. Antennas and other structures near nesting grounds interfere with flight. Thousands of Laysan albatrosses, sooty terns, and red-tailed tropicbirds died after collisions with antenna wires on Eastern Island in the mid-1960s. Lights from coastal hotels and street lamps attract and disorient fledgling seabirds on Kauai and other main islands, resulting in death for Newell's shearwaters, dark-rumped petrels, Harcourt's storm-petrels, and wedge-tailed shearwaters. On Oahu, carcasses of dead seabirds are common on both the windward and leeward approaches to the Pali and Likelike tunnels each autumn.

Aircraft frequently collide with birds when airports are constructed near their colonies. Great frigatebirds hang in huge spiraling columns above their colonies and roosting sites, posing hazards to aircraft. Some attempts to control air strikes have resulted in the loss of habitat or reductions in populations. The U.S. Coast Guard crushed more than 30,000 sooty tern eggs on Tern Island in March 1976 in an attempt to drive nesting sooties from the runway. During control programs on Sand Island between 1954 and 1964, tens of thousands of Laysan albatrosses were asphyxiated and much of the nesting habitat near the runways was destroyed. If Laysans continue to establish colonies in the main islands, control efforts near airports may become necessary for public safety.

Disturbance by Humans

Disturbance of colonies during the breeding season can reduce productivity. Eggs and nestlings of surface- and shrub-nesting species become exposed to the sun or to predators when adults are flushed from their nests. Petrels and shearwaters may be trapped or killed when their burrows are crushed underfoot, especially in the Northwestern Hawaiian Islands and Manana, where sandy soil renders burrows extremely fragile. Uncontrolled visitors to Laysan or Lisianski could cause widespread destruction of nesting burrows and bring steep population declines.

Human disturbance takes many forms. Midway was severely disrupted during World War II, when breeding grounds were used daily and nest sites of all ground- and shrub-nesting species were usurped. Except for an area around the cable compound, virtually the entire surface of Sand and Eastern was smoothed for roads, filled for underground installations, paved for runways, or covered by buildings. Many shearwaters and petrels were killed in their burrows, and those that escaped were actively persecuted when 15,000 soldiers lost sleep because of the birds' nocturnal moaning. Marines and construction men armed with two-by-fours and rods of reinforced steel clubbed thousands of albatrosses to death. Construction and Navy activities on Midway continue to disturb seabirds today, although the Navy seems to be increasingly sensitive to wildlife conservation.

The use of Kaula as a U.S. Navy and Marine Corps bomb target since 1952 has greatly disturbed a potentially important colony. Mere sonic booms can cause sooty terns to desert their colonies, and waves of explosions on nesting islands no doubt have eliminated such sensitive species as blue-gray noddies altogether. Stray bullets from marines at the gunnery range at Ulupa'u Crater in the Kaneohe Marine Corps Air Station sometimes kill red-footed boobies, which are found with bullet wounds in the chest and abdomen. Kahoolawe is used as a naval gunnery site, but except for a few tropicbirds and black noddies, it has few nesting seabirds (Table 5). Probably populations on Kahoolawe would increase if bombing were to stop.

Some unintentional disturbance stems from increasing education, affluence, and experience. Well-meaning tourists and biologists may fail to realize the damage they wreak when they enter sensitive areas. Much of such disturbance is recorded on tourists' film and in the data of biologists, who then draw questionable conclusions about poor breeding success. A single disturbance early in a nesting season may severely affect a booby or frigatebird colony. Parents may abandon nest sites, causing losses of eggs and young from predation, heat exhaustion, cold, or injury. Later in the nesting cycle disturbance may result in starvation of young that become displaced from nests. Sooty or gray-backed tern chicks that are flushed into adjacent territories may be pecked to death by neighbors. Bristle-thighed curlews or Nihoa finches may consume uncovered eggs.

The inaccessibility of most Hawaiian colonies benefits seabirds. Nevertheless, intruders venture even into the remote Northwestern Hawaiian Island colonies. The crew of an unknown U.S. Navy ship landed on Nihoa without permission in the 1960s and left a sign memorializing their visit. In the early 1960s, the Navy engaged in an unauthorized amphibious landing on Pearl and Hermes Reef and left several mementos on Southeast Island: a six-meter observation tower, fifty rusting oil drums, and tracks that persisted for years.

Introduction of Alien Species

The introduction of alien predators, insects, and vegetation to colonies is the most serious long-term threat to Hawaiian seabirds. Aliens may bring severe imbalances in simple island ecosystems. Ten of the eighteen seabirds on Christmas Island nest only on lagoon islets that are free from feral cats. After Mark Rauzon and David Woodside eliminated feral cats from Jarvis Island (a federal wildlife refuge near the equator), populations of seabirds increased dramatically there. Goats introduced on South Trinidad so denuded the island that they eliminated the nesting habitat of a huge colony of red-footed boobies. Shipwrecked pigs on Clipperton reduced vast populations of sooty terns, brown boobies, and the largest colony of masked boobies in the Pacific to a handful of nonbreeders. Dogs, mongooses, pigs, rabbits, and rats pose serious problems for Hawaiian seabirds, especially on the main islands. Dogs hunt down and kill nesting Laysan albatrosses and wedge-tailed shearwaters on Kauai, and feral pigs eat Laysans on Niihau. Mongooses have eliminated ground-nesting birds on Oahu, Hawaii, Maui, and Molokai. Rabbits overgrazed vegetation on Laysan and Lisianski during the early decades of this century, creating desert-like conditions and eliminating most nesting habitat for species that nest in shrubs and bushes.

Norwegian, black, and Polynesian rats have evolved to associate with humans. Norwegian rats arrived in Hawaii by 1838, a few decades before black rats. Polynesian rats probably arrived in the main islands with the Hawaiians some 1,500 years ago, and it seems inescapable that ancient Polynesians once visited Kure and introduced rats there. No island group reached by early Polynesians escaped colonization by Polynesian rats. Norwegian rats weigh about 400 grams; black and Polynesian rats weigh about one-third as much and are far more agile climbers. Wherever Polynesian rats are found, storm-petrels are rare or absent, even though they may be abundant on neighboring ratfree islands.

Rats are a special pest to birds on tropical oceanic islands. Natural predators are absent, there are no inclement seasons to reduce rat populations periodically, and native birds lack innate defenses against such wily interlopers. Rats do most of their work by night and can remove an egg from a nest in seconds. Individual rats may learn to prey on seabirds even when the entire population does not. Petrels are especially vulnerable because the parents leave their defenseless chicks alone in their burrows soon after they have hatched. Whenever a seabird breeding cycle coincides with a peak in rat populations or a low in food supply, rats can wreak havoc on a colony. George C. Munro recognized in 1945 the problems that rats could cause on Midway:

> A program of rodent control has been undertaken at Midway but unless rats are exterminated the Bonin petrel will be affected, Bulwer's petrels entirely killed out on the islands and the lovely white terns seriously endangered.

He was right about the first two.

Rats can do considerable damage to a colony even if each rat consumes only a few eggs a year. Because they are agile, rats can exploit a wider range of seabirds than other predators. They enter crevices and burrows in search of eggs or young, destroy nests on the ground, and may even raid cliff sites. Rats today occupy all main islands and many offshore islets, including Popoi'a, Moku'auia, and Mokoli'i. In the Northwestern Hawaiians, rats so far are restricted to Midway and Kure, where populations can reach one hundred per acre. The next introduction is merely a human error away. Refuge manager Eugene Kridler found rat poison on the wreck of a Japanese fishing vessel that broke up on the reef at Laysan in 1969. After the fishermen were rescued, they swore that no rats were aboard. Whatever the truth was, none got ashore.

Mongooses are weasel-shaped carnivores that undoubtedly have severely restricted the range of all ground-nesting birds on Oahu, Maui, Molokai, and Hawaii during the past century. They range inland up to at least 2,500 meters, so that only the highest reaches of Maui and Hawaii potentially escape their carnage. The introduction of mongooses to Kauai could have devastating effects on Newell's shearwaters. It is probably hopeless to attempt to exterminate an established mongoose population on a main island.

Introduced insects bring new diseases to isolated bird populations that have developed few immunities. Mosquitos are vectors of avian pox and malaria. An exotic mosquito on Midway is implicated in the transmission of pox, a viral disease that causes extensive facial lesions and death in albatross and red-tailed tropicbird nestlings. Pox has even cropped up on Newell's shearwaters on Kauai, probably transmitted by mosquitos from alien birds. Avian malaria, introduced to Hawaii around 1900, contributed to the demise of many endemic birds, possibly including seabirds, in the main islands. Because most declines of native bird populations occurred before the turn of the twentieth century, rats and mongooses are probably the primary culprits. Exotic insects also indirectly affect seabirds by reducing nesting habitat. Alien scales have infested and weakened Lisianski's beach magnolia, which may now be more likely to die during winter storms or drought. Many stands of beach magnolia have died off on Laysan, allowing sand movements that destabilize burrows and shrubs.

Alien vegetation may eliminate nesting habitat for ground-nesting species. Golden crown-beard, a hardy tall annual, is spreading over the central plain at Kure and eliminating breeding habitat for Laysan albatrosses and masked boobies. Other aggressive exotics on Eastern Island, Midway, have diminished open areas where masked boobies, sooty terns, and other species nest. An alien sunflower (*Bidens alba*) is common in disturbed areas of Sand Island, Midway, and provides habitat for the mosquitos that spread avian pox. Many Laysan albatrosses that nest near the sunflower stands contract pox.

Federal Laws and Policies

Governments have the resources to own and manage large tracts of important wildlife habitat. They also have the means by which to regulate the use of land they do not own provided such regulation is reasonably related to public health, safety, or promotion of the general welfare. It is inescapable that an important way to protect seabirds ashore is to acquire and manage the land they utilize. It is also expensive. The ownership of key development rights and the regulation of land use are cheaper means to the same end. The Nature Conservancy's leasing of 218 acres of Newell's shearwater nesting habitat at Kaluahonu Preserve, Kauai, indicates that conservation and management of seabird colonies are not the exclusive domain of government agencies.

Many federal laws protect Hawaiian seabirds. Since the 1960s many federal statutes have sought to ensure that wildlife conservation is considered in the federal planning process; the most important statutes are discussed here. Others, such as the Clean Air Act, the Clear Water Act (which provides for the protection of wetlands), the Comprehensive Environmental Response, Compensation and Liability Act (Superfund), and the Resource Conservation and Recovery Act, also are potentially important tools in wildlife conservation.

Federal Lands

The federal government manages seabird colonies in Hawaii in national parks and national wildlife refuges and on military bases. The federal government owns about one-tenth of the land in Hawaii, a proportion considerably smaller than in other western states. Neither the U.S. Forest Service nor the Bureau of Land Management manages any land in Hawaii. Parks are special-purpose lands, and accordingly are far less susceptible to disruption by commercial development than multiple-use lands. National park status provides the maximum protection available under federal ownership. The National Park Service Act provides that parks are to conserve wildlife to leave it "unimpaired for the enjoyment of future generations."[1] In addition, federal courts have held that the secretary of the interior has a public trust duty to protect park resources. Haleakala National Park contains most of the known dark-rumped petrel nest sites in Hawaii. The National Park Service has sponsored studies of the biology of dark-rumped petrels and has engaged in programs to protect their nest sites from predators.

The paramount federal agency for seabirds is the U.S. Fish and Wildlife Service. It manages the Hawaiian Islands National Wildlife Refuge, which includes the Northwestern Hawaiian Islands and Kilauea Point. The National Wildlife Refuge System is the only extensive system of federally owned lands that is managed chiefly for the conservation of wildlife. As multiple-use lands, however, refuges are a far cry from the inviolate sanctuaries they were originally conceived to be. The National Wildlife Refuge System Administration Act restricts the transfer, exchange, or other disposal of refuge lands.[2] The authority

[1] 16 U.S.C. § 1 (1988).
[2] 16 U.S.C. § 668dd (1988).

of the secretary of the interior to exchange lands remains unclear, but a federal court in Alaska rejected Secretary James Watt's attempt to exchange St. Matthew Island, a pristine seabird colony in the Bering Sea, for waterfowl habitat in the Yukon delta in order to allow the Atlantic Richfield Company to use St. Matthew to develop submerged oil and gas fields nearby.[3]

A refuge may be used for any activity that is compatible with the major purposes for which it was established. Unlike national parks, refuge units have rarely been designated by statute, and consequently it may be unclear why an individual refuge was established. This problem has been solved for the Hawaiian Islands National Wildlife Refuge by the development of a master plan that sets forth its goals and objectives.[4] One fundamental goal is the protection and enhancement of seabirds. Access to the fragile Northwestern Hawaiian Islands is restricted through a special-use permit system that requires approval in writing before any monitoring or research project may be undertaken.

The Hawaiian Islands National Wildlife Refuge has been under formal consideration for wilderness designation since 1969, but Congress has yet to make a decision. The U.S. Fish and Wildlife Service applies wilderness management procedures to the refuge pending Congress's decision, so that commercial activities and permanent installations are prohibited there. Seven of the large islands and atolls were designated research natural areas in 1967; the designation requires the preservation of their natural features and processes and restricts intervention to activities connected with research and education. Refuge managers have substantial discretion and authority to provide conditional-use permits for a wide variety of activities at Hawaiian refuges. It would be naive to believe that such discretion is not influenced by the prevailing political climate in the agency, in terms both of national politics and, more important, of the personalities of the current refuge managers.

The Kaneohe Marine Corps Air Station, Oahu, and Midway Naval Air Facility contain important seabird colonies. Both the Marine Corps and the Navy have elected under the Sikes Act Extension[5] to enter into cooperative agreements with the U.S. Fish and Wildlife Service and the Hawaii Department of Land and Natural Resources to carry out programs for the conservation and protection of seabirds on their bases. The U.S. Fish and Wildlife Service proposed to manage Midway as an overlay refuge in the 1970s, and in 1988 the Navy finally agreed to such an arrangement. Johnston Atoll has a similar overlay refuge status and is principally managed by the Defense Nuclear Agency. Conservation programs on military bases are always subject to the ill-defined and essentially unappealable doctrine of national security, which base commanders sometimes invoke as a means to preclude civilian scrutiny of potentially controversial activities.

[3]National Audubon Society v. Hodel, 606 F. Supp. 825 (D. Alaska 1984).

[4]U.S. Fish and Wildlife Service, *Hawaiian Islands National Wildlife Refuge Master Plan*, FES no. 86/11 (Washington, D.C., 1986).

[5]16 U.S.C. § 670a (1988).

National Environmental Policy Act

Effective wildlife conservation requires more than the acquisition of habitat and the regulation of hunting. The National Environmental Policy Act is an important statute for Hawaii's wildlife even though it mentions neither wildlife nor seabirds.[6] It ensures that each federal agency will consider the environmental impacts of its proposed actions before becoming committed to them. The effect is primarily procedural. The Supreme Court of the United States has ruled that the National Environmental Policy Act "merely prohibits uninformed—rather than unwise—agency action."[7] A court cannot substitute its judgment for that of an agency and generally defers to it when the agency has more expertise on the subject at issue than the court. A "hard look" standard of review is sometimes evoked: a court determines whether the agency took a hard look at the alternatives before reaching its decision. If an agency fails to identify the source of its facts and justify its decisions, or considers factors that it is forbidden to consider, a court does not hesitate to set the decision aside as arbitrary, capricious, or an abuse of discretion. A federal agency's interpretation of a statute that it has been entrusted to administer is given additional weight if Congress has left gaps in the statutory scheme for the agency to fill.[8]

The National Environmental Policy Act sets forth a policy to "promote efforts which will prevent or eliminate damage to the environment and the biosphere." Its critical provision is the requirement that a detailed environmental impact statement be prepared for every major federal action that significantly affects the quality of the human environment. The statement must take a systematic, interdisciplinary approach to problem solving and discuss the possible adverse effects of proposed actions and any reasonable alternatives. Alternatives to proposed actions must be included and any irrevocable commitments of resources, such as the use of wildlife habitat, must be identified.

The National Environmental Policy Act has enlarged the authority of non-resource agencies, including the Department of Defense, to encompass environmental protection. It requires resource agencies such as the U.S. Fish and Wildlife Service and the National Marine Fisheries Service to review the environmental decisions of other federal agencies. One hidden benefit has been the elimination of the worst projects (such projects are no longer formally proposed) and the omission of objectionable portions of others, which are changed during the planning process to mitigate adverse environmental consequences. Its notice requirements allow interested citizens and organizations to comment on draft environmental impact statements, thereby allowing wide participation and increased expertise in planning. In many cases the need to prepare an environmental impact statement brings individuals and organizations together for cosponsorship to achieve a consensus in regard to priorities.

[6] 42 U.S.C. §§ 4321–47 (1988).
[7] Robertson v. Methow Valley Citizens Council, 57 U.S.L.W. 4497, 4502 (1989).
[8] Chevron U.S.A., Inc. v. Natural Resources Defense Council, 467 U.S. 837 (1984).

Any major federal action that might have a detrimental effect on a seabird colony would be scrutinized in an environmental impact statement. Conservationists and biologists would have an opportunity to propose alternatives that might mitigate or eliminate damage or to persuade an agency to abandon the project. Environmental impact statements are required for ongoing programs as well as for specific projects. The U.S. Fish and Wildlife Service prepared a statement for the Operation of the National Wildlife Refuge System in 1976 and planned to update it by 1991. By its nature the National Environmental Policy Act provides some judicial review and public intervention in the administration of wildlife programs.

A major failing of the National Environmental Policy Act is an absence of any requirement to prepare retrospective evaluations of past environmental impact statements to compare predicted with actual consequences. Although some agencies might be embarrassed by the revelation of past errors, the information obtained would vastly improve predictive techniques for future projects.

Migratory Bird Treaty Act

The Migratory Bird Treaty Act was the first federal attempt to limit the killing or taking of migratory birds.[9] Although the impetus for its passage was the desire to solve problems with such species as ducks and geese, it established a year-round "close season" for all migratory birds except game birds. Seabirds and their nests and eggs cannot be taken, killed, or sold without a permit. The Migratory Bird Treaty Act implements treaties with Great Britain (on behalf of Canada), Mexico, Japan, and the USSR and requires the secretary of the interior to protect the birds listed in the four conventions. By regulation, all of the twenty-two species of seabirds that nest in Hawaii are protected, notwithstanding the fact that several are nonmigratory in the strict biological sense.[10] The treaty with Japan is unusual in requiring protection of the ecological balance of unique island environments. The U.S.–Soviet treaty prohibits the "disturbance of nesting colonies" and directs both nations to establish and manage preserves. Despite such provisions in the recent treaties with Japan and the USSR, the Migratory Bird Treaty Act remains very similar to the original 1918 act and has not been amended to include the additional features. Whether the treaties are self-executing and thus do not require implementing domestic legislation is an open question.

The U.S. Fish and Wildlife Service administers the Migratory Bird Treaty Act on both private and public land. Scientists who wish to collect seabirds as part of a research program must first obtain a permit. Because "taking" is broadly defined by regulation, the agency has fairly wide authority to protect seabirds; the regulation should, among other things, prohibit disturbance to breeding

[9] 16 U.S.C. §§ 703–12 (1988).
[10] 50 C.F.R. § 10.13 (1989).

colonies and accidental killing by pesticide sprays.[11] Hawaii is administered by the U.S. Fish and Wildlife Service's regional office in Portland, Oregon. Because the region encompasses six states, including California, Hawaii's problems are often lost in the shuffle. The regional office has established a marine bird policy for all refuges, which includes the goals of maintaining the population levels of all marine birds and removing all introduced predators from their colonies.

Endangered Species Act

The Endangered Species Act provides a powerful weapon to protect species that the federal government has designated as endangered or threatened.[12] All federal agencies, not just those whose mandate is conservation, must carry out programs to conserve listed species. Loss of habitat is the root cause of the endangered status of most island-dwelling creatures, and many activities that cause such losses are directly undertaken or indirectly authorized by federal agencies. Such activities fall within the ambit of the statute. Although the Endangered Species Act also provides for the designation of critical habitat (areas essential to the conservation of a species), critical habitat has not been designated for any Hawaiian seabird. Dark-rumped petrels and short-tailed albatrosses are listed as endangered species[13] because they are in danger of extinction throughout all or a significant portion of their ranges. Newell's shearwaters are listed as threatened because they are likely to become endangered within the foreseeable future. As a threatened species, Newell's could be protected by "regulations that are necessary and advisable" to provide for their conservation, but no such regulations have been issued. Region One of the U.S. Fish and Wildlife Service has informally designated sooty storm-petrels as a sensitive species and monitors their status for possible protection under the Endangered Species Act. The U.S. Fish and Wildlife Service has issued a recovery plan for dark-rumped shearwaters and Newell's shearwaters, but unfortunately it was not accompanied by an environmental impact statement.[14]

Dark-rumps and short-tails are fully protected by the act and cannot be sold or taken. The statute broadly defines "taking" to encompass harassment, harm, pursuit, capture, collection, shooting, and killing. Just as important, the regulations define "harm" to include significant modification or degradation of habitat. As the State of Hawaii has twice learned to its chagrin in *Palila* v. *Department of Land and Natural Resources*, state actions that degrade the feeding, roosting, or nesting habitat of an endangered species can be enjoined by a federal court.[15] The Endangered Species Act allows taking if it is incidental to the

[11] 50 C.F.R. § 10.12 (1989).
[12] 16 U.S.C. §§ 1531–44 (1988).
[13] 50 C.F.R. § 17.11 (1989).
[14] U.S. Fish and Wildlife Service, *Hawaiian Dark-Rumped Petrel and Newell's Shearwater Recovery Plan* (Portland, Ore., 1983).
[15] Palila v. Hawaii Department of Land and Natural Resources (DLNR), 471 F. Supp. 985 (D. Hawaii 1979), *aff'd*, 639 F.2d 495 (9th Cir. 1981); Palila v. DLNR, 649 F. Supp. 1070 (D. Hawaii 1986), *aff'd*, 852 F.2d 1106 (9th Cir. 1988).

carrying out of an otherwise lawful activity such as forestry, clearing land, or fishing. However, no incidental take permit may be issued without the submission of a conservation plan, which must include means to mitigate the harm caused by the taking of an endangered species.

There is a great deal of confusion concerning the importance of designating critical habitat. The designation has operative significance under the Endangered Species Act only through a process by which each federal agency consults with the U.S. Fish and Wildlife Service or the National Marine Fisheries Service to ensure that any action the agency authorizes, funds, or carries out does not jeopardize the continued existence of any endangered or threatened species or result in the destruction of critical habitat. Because the destruction of critical habitat would necessarily place any endangered species in jeopardy, the formal designation of critical habitat does not seem to have profound legal significance; yet it has practical significance. Critical habitat identifies for agencies, judges, and corporate executives the precise locations where activities will definitely run afoul of the Endangered Species Act. The location of a new highway, airport, or military installation is much more likely to be changed at an early stage in the planning process, before large resources have been committed to a particular plan, when the proposed site has been designated a critical habitat. The statute is so broadly construed that even federal loan guarantees may trigger the process of consultation with the U.S. Fish and Wildlife Service or the National Marine Fisheries Service. If either agency issues a biological opinion that a proposed action jeopardizes an endangered or threatened species, the sponsoring agency must adopt reasonable and prudent alternatives to avoid the adverse effects.

Most of the known dark-rump nests in Hawaii are located within Haleakala National Park, Maui. Though it is unlikely that the National Park Service would allow activities that would directly jeopardize the dark-rumps' nesting habitat, the Endangered Species Act requires extensive review and public comment if such actions should be proposed. It also allows citizens' suits to compel the agency to comply with the statute's requirements. Short-tails come ashore occasionally on federal lands in the Northwestern Hawaiian Islands but probably do not face any immediate threats except possibly harassment on Midway. Newell's shearwater nests in the mountains of Kauai are threatened primarily by introduced predators that need no permit to enter a Newell's colony. Hence the Endangered Species Act has limited value on land to Hawaiian seabirds today. If new dark-rumped nest sites were discovered, however, the act's critical habitat provisions could be employed to provide additional protection.

State of Hawaii Laws and Policies

Seabirds are potentially protected by a wide variety of Hawaii statutes, many of which roughly parallel those of the federal government. The State of Hawaii also exerts powerful control of land development through its land use planning statutes, which, when properly applied, provide substantial protection to all wildlife. Some of the most pertinent of the many state statutes and policies are considered here.

State Lands

The state is the largest landowner in Hawaii, holding over a third of the land. The federal government turned over most of its land to Hawaii when it became a state in 1959, retaining primarily military land, the national parks, and the refuges. The state is responsible for seabird colonies in wildlife sanctuaries and forest reserves. None of its seventy-four parks or eighteen natural area reserves (communities of natural flora and fauna) is known to contain a seabird colony. However, Ka'ena Point Natural Area Reserve on Oahu has roosting Laysan albatrosses that need protection, and dark-rumped petrels may nest in the Hono O Na Pali Natural Area Reserve on Kauai. Over the objections of some local fishermen, the Board of Land and Natural Resources has designated thirty-nine islets, islands, and rocks as the Hawaii State Seabird Sanctuary to conserve, manage, and protect seabirds.[16] It is prohibited to remove, disturb, injure, kill, or possess any seabird at any sanctuary. The Department of Land and Natural Resources does not generally restrict access to the sanctuary, but a permit is required for Moku Manu, Manana, Mokuho'oniki, and Kure.

The state manages vast amounts of forest reserves in the mountainous interiors of the main islands. Within these holdings are colonies of Newell's shearwaters, Harcourt's storm-petrels, white-tailed tropicbirds, and dark-rumped petrels. Owing in part to difficulties in obtaining access and consequent lack of knowledge, the Department of Land and Natural Resources does little if any active seabird management in its forest reserves.

Statewide Land Use Planning

The State of Hawaii intensely regulates the use of land and has the most comprehensive land use controls in the United States. Unlike any other state, Hawaii has enacted its plan as a statute.[17] Among the state plan's many objectives are the exercise of an overall conservation ethic in the use of Hawaii's natural resources and effective protection of Hawaii's unique and fragile environmental resources. The state plan provides broad guidelines for state activity, but its use is limited because it also contains a long list of economic and development goals that inherently conflict with environmental goals without providing any means of reconciliation.

The Hawaii Land Use Act zones all public and private land in the state into

[16]Title 13, DLNR, subtitle 5, Forestry and Wildlife; pt. 2, chap. 125 (Sanctuaries) (September 23, 1981).

[17]Hawaii Rev. Stat. § 226 (1985).

four land use districts: urban, rural, agricultural, and conservation.[18] Land can be reclassified by application to the Land Use Commission, but any reclassification must conform to the state plan. Conservation land is further classified into four subzones of descending degrees of protection: protective, limited, resource, and general. The six-member Board of Land and Natural Resources has general powers to oversee all land held by the state and specific authority to grant or deny applications for the use of conservation land, which accounts for almost half of the state's land. The members of this powerful board are appointed by the governor primarily on the basis of political connections rather than any expertise in the management of natural resources.

Following the enactment by Congress of the Coastal Zone Management Act[19] of 1972, Hawaii adopted the Hawaii Coastal Zone Management Act[20] as a comprehensive regulatory scheme to protect the environment and resources of Hawaii's coastal areas. Land along the coast has faced and continues to face the most intensive development pressures. The state's declared policy is to preserve, protect, and, where possible, restore the natural resources of Hawaii's coastal zone. Most seabird colonies are located in the coastal zone and fall within the ambit of the statute's protections.

The implementation of the Hawaii Coastal Zone Management Act is delegated largely to the four counties that administer the procedures governing the issuance of permits to develop land within "special management areas," defined as the first hundred yards landward of the shoreline. The counties are required to designate as special management areas locations that require extraordinary protection, including essential habitat for wildlife, on county maps. No major development may be approved within a special management area unless the county finds it will have no substantial adverse environmental or ecological effect, except when such adverse effects are clearly outweighed by considerations of public health and safety. A permit to use a special management area may not be granted unless a public hearing is held at which interested individuals may testify. It seems unlikely that any county could find that a proposed development would cause no substantial adverse environmental effect if it imperiled one of the few remaining seabird colonies in the main islands. The Hawaii Coastal Zone Management Act grants broad standing to any person to challenge in court a county's findings or its compliance with the statute's objectives.

The City and County of Honolulu, in which many of Hawaii's coastal seabird colonies are located, has prepared a development plan for the Northwestern Hawaiian Islands and the islands of windward Oahu. It seeks to preserve and protect the environment and wildlife of those islands, emphasizing protection of the resources together with controlled use for educational, research, and

[18]Ibid., § 205.
[19]16 U.S.C. §§ 1451–64 (1988).
[20]Hawaii Rev. Stat. § 205A (1985).

recreational purposes. There would be extensive input by the public and oversight by the City Council before any special management area permit could be issued for a proposed development at any seabird colony. As a practical matter, the land use control laws in Hawaii make it extremely difficult for detrimental projects at coastal seabird colonies to be approved.

Hawaii Environmental Policy Act

The Hawaii Environmental Policy Act generally parallels its federal counterpart and addresses major activities that have state but no federal involvement. Proposed actions that fall within its scope include the use of state and county lands and funds, and any action that requires a permit for the use of a special management area or conservation district.[21] The responsible agency for such actions must at a minimum prepare an environmental assessment to determine whether a full-blown statement is necessary. An environmental impact statement is required for actions that would have a significant effect on the environment, including those that are contrary to Hawaii's environmental policies or would irrevocably commit natural resources. No project may go forward until the statement has been accepted.

Whenever a statement is required under both federal and state environmental policy acts, federal and state agencies cooperate to reduce duplication and frequently issue joint statements. Like the federal statute, the Hawaii Environmental Policy Act is primarily procedural, and its chief virtue is the public examination of the adverse environmental effects of proposed projects. It cannot by itself prevent the state government from undertaking an environmentally disastrous project, but it can be used to halt a project for procedural irregularities. Any state project that would directly harm a seabird colony would probably require an environmental impact statement. Interested individuals, organizations, and agencies would have an opportunity to suggest ways to mitigate the project's worst effects or oppose its approval altogether.

Hawaii Wildlife Protection Statutes and Policies

The State of Hawaii has enacted a variety of statutes to protect all wildlife. The legislature has declared that all indigenous wildlife is an integral part of Hawaii's native ecosystems and is part of the living heritage of Hawaii. It is unlawful for any person to take, possess, sell, or transport any nondomesticated member of the animal kingdom except pursuant to regulations issued by the Department of Land and Natural Resources. By regulation, the department may issue permits only for scientific or educational purposes for which collecting is essential.

The Hawaii Endangered Species Act provides some additional protection to

[21]Hawaii Rev. Stat. § 343 (1985).

any species listed as endangered or threatened by its federal counterpart or any indigenous species that the Department of Land and Natural Resources has listed because its continued existence is jeopardized by destruction of habitat, disease, or similar factors.[22] In addition to the seabirds that the federal government has listed as endangered or threatened, the department has designated as endangered white terns on Oahu and Harcourt's storm-petrels throughout their range in Hawaii. The Hawaii statute focuses exclusively on prohibiting the direct taking of endangered species. It lacks provisions to protect habitat or to command state agencies other than the Department of Land and Natural Resources to conserve and protect endangered or threatened species. Hence it is simpler and far less useful than the federal act.

The Department of Land and Natural Resources set forth its seabird policies in its 1984 wildlife plan.[23] The plan assigns its highest priority to endangered and threatened seabirds and species that are important to the fishing industry. The department seeks to remove or control the factors that limit the survival of threatened and endangered seabirds. Although the plan suggests that all seabird colonies be monitored at least once each year, the department has never done a single survey on many of the islands in the state seabird sanctuary. The plan also advocates the removal of rodents from seabird sanctuaries and encourages human use except where visitors are detrimental to seabirds or their habitats.

Efficacy

How well do the laws work? Statutes that establish preserves and outlaw taking have aided the remarkable recovery of populations that seemed destined for extinction not so many decades ago. Yet the mere designation of sanctuaries and control of direct use of seabirds leaves many important terrestrial conservation problems unaddressed. Few Hawaiian seabirds today are intentionally killed, and no commercial enterprise consumes any part of a seabird.

The key to any statutory framework that protects wildlife is enforcement. Here the record is all too often poor. The Northwestern Hawaiian Islands are difficult to patrol and seem almost as remote today as they were at the turn of the twentieth century. Fortunately the isolation that hampers enforcement also protects the wildlife. Less understandable is the poor enforcement at the state seabird sanctuary. A single patrol boat could effectively monitor visitors to the colonies offshore Oahu, and occasional inspections at islets offshore Kauai, Maui, Molokai, Lanai, and Hawaii are warranted. The costs of such patrols would not be large. Their absence is a result of the general inattention of the Department of Land and Natural Resources to the conservation of marine resources.

[22]Ibid., § 195D.

[23]State of Hawaii, Department of Land and Natural Resources, *Hawaii Wildlife Plan* (Honolulu, 1984).

The destruction of nesting habitat is a major threat to Hawaiian seabirds, and current laws and policies are mixed in this regard. State and federal agencies have been successful in designating large and conspicuous seabird colonies as parks, refuges, or sanctuaries. Undoubtedly small colonies in the forested mountains of the outer islands need similar protection from logging, grazing, agriculture, and other human activities, but they have not been located. Where passive management is sufficient, seabirds are protected. Many colonies, however, need intense and active management programs to control or eliminate mongooses, rats, and dogs. The elimination of rats from seabird colonies, especially the islands offshore Oahu, has received insufficient attention.

Agency wildlife programs are only as strong as their funding. They vary with the interests of the personnel involved, especially on military bases, where programs may change radically to accommodate the interests of an individual base commander. Committed individuals who work in the wildlife agencies cannot be faulted for the low funding levels and inadequate staffing that hamper their best efforts. In an era when the federal government is generally cutting back its services, the state has not signaled any clear intention to pick up the slack. The *Hawaii Wildlife Plan* is an eloquent statement of policy, yet like so many state statutes, plans, and policies, it is poorly implemented. It includes no funding levels and no dates by which the Department of Land and Natural Resources must reach its stated goals. No wonder that progress in meeting its goals is limited.

17 CONSERVATION AT SEA

The downy Laysan albatross had the misfortune of being fed as I passed by its nest on Green Island, Kure. It would lose its latest meal to science. As I held the wriggling bird firmly and turned it upside down, I knew that I was taking merely a single feeding. Some of my colleagues elsewhere routinely disemboweled seabirds to collect food samples. While the nestling struggled and kicked in vain, it retched to the sand a bouillabaisse of squid mantles, semidigested fish flesh, and fish eggs marinated in stomach oil. When I set the young bird upright again, it clacked its bill at me and attempted to regain its lost dignity. As I turned my head to avoid the putrid effluvia while scooping the sample into a jar, I discovered a unique item—a brown plastic buffalo the size of a kukui nut. Floating wastes of late-twentieth-century civilization such as toothpaste caps, balloons, tiny toy tyrannosaurs, and plastic fibers have been found in the stomachs of at least sixteen Hawaiian seabird species. The albatross had swallowed the symbol of the U.S. Department of the Interior, parent agency of the U.S. Fish and Wildlife Service. Hawaiian seabirds face many conservation problems in the ocean, but the U.S. Fish and Wildlife Service has little authority to help.

The ocean is the home and the larder of seabirds, and it is no surprise that they encounter conservation problems there. Because few marine biologists have considered seabirds as components of marine ecosystems and few ornithologists have also been oceanographers, our knowledge of seabirds at sea pales in comparison with what we know of their natural history ashore. Humans have not yet developed techniques to exploit the ocean's resources as efficiently as they do those of the land. Consequently, conservation principles and laws to protect the ocean environment are less developed.

Threats at Sea

Most seaward conservation problems emanate from fisheries or pollution. Seabirds can be directly injured by fishing gear and indirectly harmed by decreases in the availability of prey. A few masked boobies are hooked during trolling operations or tangled in fishing nets near the Hawaiian Islands. Black-footed and Laysan albatrosses die when they become snagged or netted in the Japanese long-line tuna fishery farther north, but such losses seem to be minor. A potentially serious problem is the high-seas drift gillnet fishery for squid operated by Japan, Taiwan, and South Korea some 2,000 kilometers north of Hawaii. The incidental effects of such oceanic strip mining on marine wildlife are poorly studied. Drift gillnets, manufactured from finely spun plastic strands, are designed to entrap any marine creature that cannot pass through their narrow mesh. Floats support the nets on the surface and weights hold down the bottom of the fifteen-meter-deep "curtains of death." The monofilament nets in use extend distances of fifteen kilometers and are not biodegradable. Lost or discarded nets lurk beneath the surface for years until they finally sink or are destroyed. Since 1981, the equivalent of 1.5 million kilometers of drift gillnet has been set for squid each year near the North Pacific subtropical convergence, a far greater length of net than at any other fishery on earth. The England-based International Council for Bird Preservation estimates that one million birds drown each year, possibly including endangered short-tailed albatrosses and dark-rumped petrels. The most vulnerable seabirds are those that feed beneath the sea by diving or pursuit plunging, but scavengers such as albatrosses become entangled when they attempt to eat entrapped organisms. Of less importance to Hawaiian seabirds, a drift gillnet fishery for tunas began in the South Pacific during the late 1980s.

Decreases in seabird populations resulting from changes in the food web have been observed the world over. Biologists have long acknowledged the relationship between the reproductive success of seabirds and the availability of food. As apex predators in the marine ecosystem, seabirds are potential competitors of certain commercial fisheries. Overfishing for anchovies off Peru by an overcapitalized fishery, combined with the effects of El Niño, has reduced the seabird population there by 90 percent since the mid-1950s. In Hawaii, unmanaged yellowfin and skipjack tuna fisheries could threaten seabirds because many species rely on tunas to drive their prey to the surface. If too many surface-feeding tunas were fished from Hawaiian waters, the number of birds that the sea could sustain would diminish. Seabirds are most vulnerable during their breeding seasons, especially from March to September. Many species migrate or feed far out to sea during fall and winter, and would be less affected then by fisheries near the islands.

The Hawaii pole-and-line skipjack tuna fishery requires a reliable supply of good baitfish. Shoaling schools of small fishes such as anchovies and herrings and juvenile forms of mackerel scad, big-eyed scad, and goatfish are sometimes used for bait. Such fish are also important components of the diets of terns, shearwaters, and boobies. Adult mackerel scad are commercially valuable and constitute a substantial proportion of the diet of boobies, tropicbirds, and frigatebirds. Squid accounts for over half the prey taken by Hawaiian seabirds, and although techniques to catch squid in Hawaiian waters are rudimentary, if such a fishery were established it could create ecological imbalances that would affect seabirds. Depletion of any major prey organism near a breeding colony would probably result in the starvation of young, abandonment of breeding, and, if shortages were severe, starvation of adults.

Oil, plastics, and agricultural pesticides escape into Hawaiian waters from ships and coastal activities. Oil enters the ocean from transfer operations, shipwrecks, and bilge discharge. In 1977 the *Irenes Challenge* broke apart some eighty kilometers north of Lisianski and spilled over five million gallons of crude oil, about half the quantity released into Prince William Sound, Alaska, by the *Exxon Valdez*. Fortunately none came ashore and Lisianski was not "exxoned." In the past two decades several fishing vessels have grounded at the Northwestern Hawaiian Islands and a Greek freighter smashed into the reef at French Frigate Shoals. Oil clogs the plumage of seabirds and destroys insulation, killing birds through chilling and stress, even in tropical waters. Michael Fry's studies of wedge-tailed shearwaters on Manana indicate that their reproductive success decreases sharply after they have ingested oil or been exposed to it externally. Sooty terns, white terns, black noddies, masked boobies, red-footed boobies, and Laysan albatrosses have been observed covered with oil in Hawaii.

The finding of plastic, styrofoam, and other flotsam in seabirds' stomachs is a relatively recent phenomenon; the first such articles appeared in the 1960s with the advent of styrofoam and longer-lived, buoyant materials made from synthetic fibers. More than eighty seabird species throughout the world are known to ingest plastic. Most such material in Hawaiian waters originates in Japan. It is hoped that the decision by Japan's Environment Agency in 1988 to ban ocean dumping of all plastic waste will eventually diminish this problem. Pumice and other floating objects were found in albatrosses' stomachs during the 1940s and 1950s, but I have located no record of their ingestion of industrial products. Paul R. Sievert has discovered that albatrosses ingest plastic more frequently than other seabirds do and that they favor tan-colored items. Terns rarely consume plastic but shearwaters, Bonin petrels, sooty storm-petrels, great frigatebirds, and red-tailed tropicbirds do so fairly commonly.

Sublethal doses of toxic chemicals can be ingested with food and build up in the body fat of seabirds. Such materials tend to be released when fat is mobilized during periods of stress. Accordingly, toxic materials may contribute to death when seabirds encounter bad weather or food shortages. The eggs of sooty terns, wedge-tailed shearwaters, and red-footed boobies in the main islands, French Frigate Shoals, Laysan, and Midway contain chlorinated hydrocarbons such as PCBs, DDE, and DDT and heavy metals such as mercury. Chlorinated hydrocarbons, which cause eggshells to become thin, may emanate from agricultural runoff, industrial wastes, and possibly plastics. Mercury probably enters the Hawaiian environment through volcanic activity. The visceral fat of albatrosses has appreciable residues of DDT, DDE, and PCBs, but as yet no measurable eggshell thinning has been found.

Seabirds may someday face conservation problems from industrial uses of the Pacific Ocean. Ocean thermal energy conversion (OTEC), which harnesses the energy released when warm surface water comes in contact with cold deep water, has been touted as a cheap means of generating electricity in Hawaii. A 40-megawatt closed-cycle plant planned for the waters just offshore Kahe Point, Oahu, could be the world's first large-scale plant of this kind. Upon completion, it would supply one-twentieth of Honolulu's electricity. Ocean engineers envision a day when floating plants will range over the open ocean to generate and store electricity. Such activities could bring massive changes to surface sea temperatures offshore and affect seabirds and other marine organisms in ways that cannot yet be imagined. Deep-sea mining for manganese nodules on the abyssal plain and on the cobalt-rich manganese crusts of submerged island slopes and seamounts surrounding the Hawaiian Islands is in the planning stages. The federal and state governments have jointly issued an environmental impact statement concerning exploration permits for the seabed near Hawaii. Any mining operation that hauled thousands of tons of ore through the water column to the surface could ultimately increase concentrations of heavy metals in seabirds.

Jurisdiction

A review of the boundaries of state, federal, and international waters and an appreciation of the distinctions that make the seabed and the water column separate legal jurisdictions are prerequisites for understanding the protection and management of the marine environment. The seaward limits of national jurisdiction have been in flux since the 1950s. The Law of the Sea Treaty has been signed by most of the nations of the world and should soon enter into force. Despite the fact that several important maritime nations (including the United States) are not parties to it, the treaty has already greatly changed customary international law.

The State of Hawaii has authority over the submerged land and the water column out to three nautical miles. The Submerged Lands Act in 1953 transferred such authority to the states to negate the U.S. Supreme Court's decision that the federal government owns all rights seaward of the coast.[1] Congress enacted the Outer Continental Shelf Lands Act at the same time to authorize the Department of the Interior to exercise primary control beyond three nautical miles, including leasing and management of the resources of the outer

[1] 43 U.S.C. §§ 1301–15 (1988).

continental shelf.[2] Despite Hawaii's lack of a geological continental shelf, Congress designated a legal one. Within three miles of its coastline, the state has broad authority to regulate marine activities. Hawaii's constitution authorizes the management and control of marine, seabed, and other resources located within the state's boundaries and provides that state lands are held in trust for the benefit of the people. The environmental provisions of the state plan discussed in chapter 16 apply equally to the ocean.

The seaward extent of federal jurisdiction surrounding the Hawaiian Islands is a question of international law. President Reagan proclaimed a 200-nautical-mile exclusive economic zone in 1983 around the United States of America and its overseas territories and possessions, encompassing 2.5 million square miles (Figure 16).[3] In 1988, President Reagan proclaimed a twelve-nautical-mile territorial sea, extending the prior claim of the United States by nine nautical miles.[4] The legal effects of the proclamations are somewhat uncertain and may require implementing legislation by Congress. Although the United States has not signed the Law of the Sea Treaty, claims for 200-nautical-mile exclusive economic zones and twelve-nautical-mile territorial seas surrounding coastal nations have become so widespread that they are now established principles of customary international law. President Reagan's 1988 proclamation emphasized that an extension of the territorial sea does not affect the three-mile jurisdiction of states such as Hawaii. Waters seaward of 200 nautical miles are international.

Marine Preserves

Marine preserves, like parks and sanctuaries on land, are one means to protect marine wildlife. The state conservation functional plan requires the Department of Land and Natural Resources to identify and maintain a comprehensive inventory of critical environmental areas and to establish sanctuaries when necessary to protect critical habitats of endangered species. Wildlife habitats in state waters are included in this mandate. The natural area reserves program[5] could provide protection for marine areas, but all eighteen of the current reserves are terrestrial. The seven marine life conservation districts were established primarily to regulate fishing on nearshore reefs. The Hawaii Ocean and Submerged Leasing Act provides additional protection by restricting leases in state waters where a marine life conservation management area program or a natural area reserve would suffer adverse effects.

The federal government can designate marine waters as national parks, wildlife refuges, estuarine sanctuaries, or marine sanctuaries. Although several

[2]Ibid., §§ 1331–56 (1988).

[3]Exclusive Economic Zone of the United States of America, Proc. no. 5030, 48 Fed. Reg. 10605 (1983).

[4]Territorial Sea of the United States of America, Proc. no. 5928, 54 Fed. Reg. 777 (1989).

[5]Hawaii Rev. Stat. § 195 (1985).

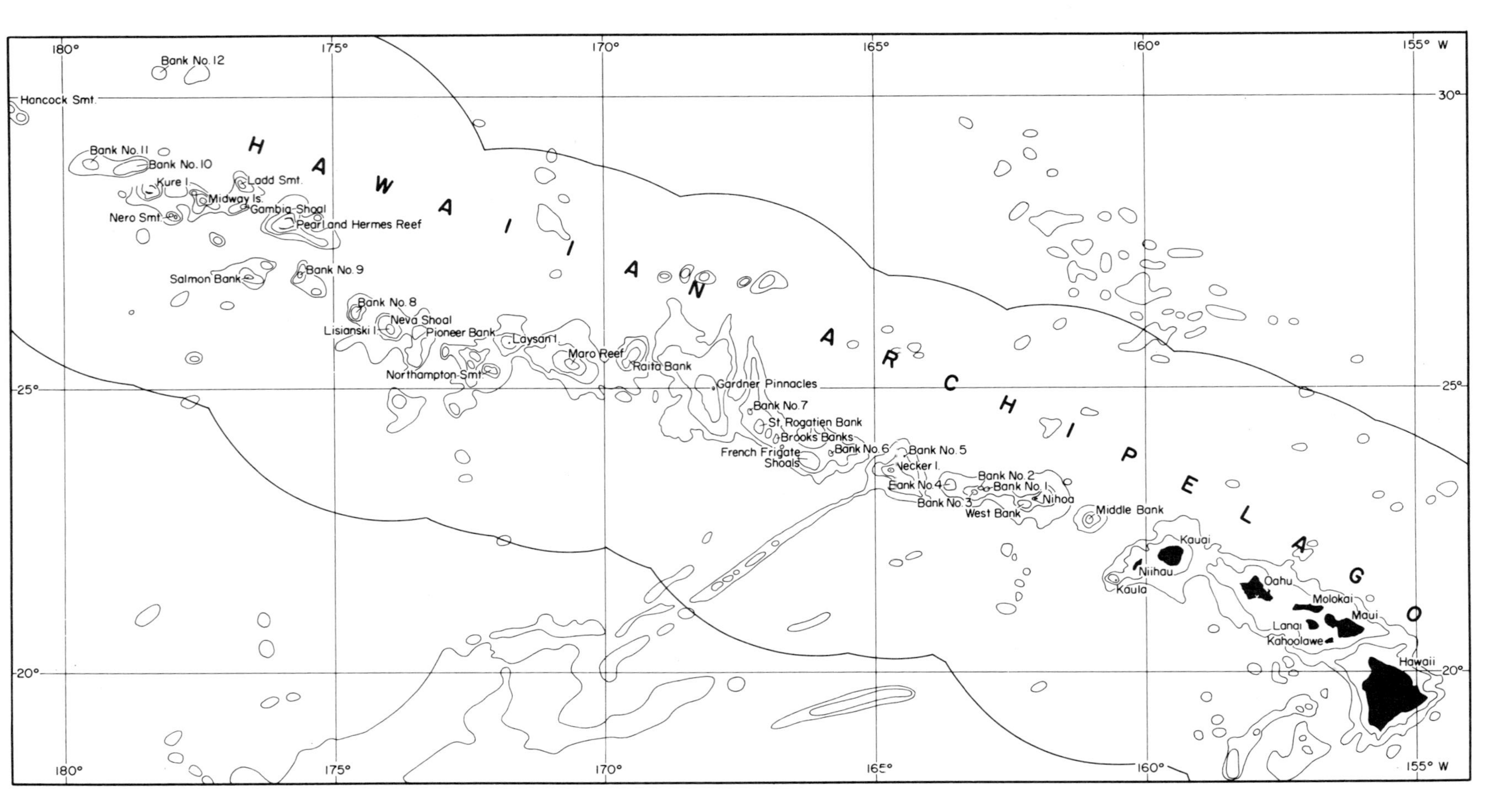

Figure 16. The exclusive economic zone around Hawaii

national parks in Hawaii border the ocean, none includes marine waters. The lagoons at French Frigate Shoals, Pearl and Hermes Reef, and Maro Reef and certain waters surrounding Laysan are within the boundaries of the Hawaiian Islands National Wildlife Refuge. The estuary at Waimanu Valley, Hawaii, has been designated an estuarine sanctuary and is administered by the state under federal guidelines.

The national marine sanctuary program is a comprehensive federal-state marine management effort that fosters multiple use and provides environmental protection that otherwise might be unavailable. Regulations prohibiting sanctuary violations can have real teeth. In 1986 the owners of the M/V *Wellwood* agreed to pay the federal government more than $6 million for damages caused when the freighter ran aground in the Key Largo National Marine Sanctuary. The National Ocean and Atmospheric Administration provides overall management of the program, but a site may be managed by a state agency. None of the nation's eleven existing marine sanctuaries or any currently proposed sanctuary is located in Hawaii. Governor George Ariyoshi exercised his veto authority over a proposed humpbacked whale sanctuary in the waters between Maui, Molokai, and Lanai in 1984 after objections by some "commercial" fishermen, many of whom cannot make a profit without generous state and federal tax subsidies. The veto decision was particularly unfortunate because the tremendous economic benefits that accrue from whale-watching tours apparently were not considered. The Marine Sanctuary Act was subsequently amended to require consultation with the affected regional fishery council and Congress when fishing regulations are proposed within a marine sanctuary.[6] No Hawaii site is included in the National Ocean and Atmospheric Administration's current site evaluation list, in part because no public meeting was held in Hawaii during the nomination process. It seems unlikely that the federal agency will expend its limited resources to propose another marine sanctuary in Hawaii unless the state initiates the proposal, as California did for its marine sanctuaries in the Channel Islands and the Gulf of Farallon.

The Intergovernmental Maritime Organization, an agency of the United Nations, has declared a 50-mile radius surrounding the islands and atolls between Nihoa and Pearl and Hermes Reef as an "area to be avoided." The designation is merely advisory and applies only to ships of 1,000 gross tons carrying oil or hazardous chemicals.

Industrial Uses of the Ocean

OTEC Plants

The Ocean Thermal Energy Conversion Act of 1980 established one-stop licensing for ocean thermal energy conversion (OTEC) plants and effectively

[6]16 U.S.C. §§ 1431–45 (1988).

provides for federal rather than state regulation for any such plant under United States control, wherever it may be located.[7] No such plant may be operated without a permit. The intent of the act is to protect the marine and coastal environment and to prevent or minimize any adverse effects from OTEC plants. The National Ocean and Atmospheric Administration is the lead agency that coordinates with interested federal, state, and county agencies through an environmental impact statement process. A governor can effectively veto any proposed license for an OTEC plant in state waters. The federal Environmental Protection Agency must certify compliance with laws over which it has authority, including the thermal discharge provisions of the Clean Water Act. A licensee of an operating plant must monitor the environmental effects of its facility and submit its data to the National Ocean and Atmospheric Administration, which is empowered to terminate operations if the plant poses an imminent and substantial threat to the environment. The environmental effects of this new technology have been extensively investigated and a programmatic environmental impact statement has been issued. Baseline studies continue, including efforts to determine the cumulative effects of the operation of such plants.

Floating OTEC plants that are not subject to stringent environmental controls could be operated 200 or more nautical miles offshore Hawaii by citizens of countries other than the United States. Although the law is somewhat unclear, the Law of the Sea Treaty and customary international law seem to require the equivalent of an environmental impact statement for the operation of such plants.

Deep-Sea Mining

Deep-sea mining within the United States' exclusive economic zone is regulated by the Department of the Interior pursuant to the Outer Continental Shelf Lands Act. Mining more than 200 nautical miles offshore is authorized by the National Ocean and Atmospheric Administration pursuant to the Deep Seabed Hard Mineral Resources Act.[8] The latter was passed in 1980 in response to frustration over the progress of the negotiations over the Law of the Sea Treaty. A permit under either statute is considered to be a major federal action under the National Environmental Policy Act, which requires an environmental impact statement based on baseline and monitoring studies. Both statutes allow citizens' suits to compel compliance with the provisions of a mining lease or federal regulations. Under the Outer Continental Shelf Lands Act, each of three major stages—leasing, exploration, and development and production—requires a separate environmental impact statement. The secretary of the interior is responsible for conserving marine life and must coordinate and consult with

[7] 42 U.S.C. §§ 9101–68 (1988).
[8] 30 U.S.C. §§ 1401–73 (1988).

state and local governments and any affected fishery management council. The secretary is required to consider available relevant environmental information in developing regulations and conditions for permits.

The Deep Seabed Hard Mineral Resources Act raises a stiff barrier to the issuance of a permit: no permit may be issued for exploration or commercial recovery unless the National Ocean and Atmospheric Administration determines that the activity cannot reasonably be expected to have a significant adverse effect on the quality of the environment. Once a license or permit is granted, it must be modifed by the agency if new information indicates that changes are necessary to protect the environment. Each license and permit authorizes federal observers and requires environmental monitoring of the mine site. A permit can be suspended or revoked to prevent significant adverse effects or for noncompliance with its terms and conditions.

Marine Debris

In 1987 Congress enacted the Marine Plastic Pollution Research and Control Act to address the problems that marine debris causes wildlife.[9] It prohibits the dumping of "garbage"—presumably including paper, glass, and metal in addition to plastic—within the 200-mile exclusive economic zone. U.S. ports must provide disposal sites for garbage, ships must keep records concerning their garbage, and the U.S. Coast Guard is empowered to inspect ships at sea. The federal government is instituting an education program for fishermen, recreational boaters, and industries about the effects of plastics and other debris in the ocean. Ultimately education and improved technology are the keys to this problem's solution. Until shipping companies, fishermen, weekend sailors, commercial operators, and especially the U.S. Navy understand the hazards of marine debris, little genuine progress seems likely. Japan's Ministry of International Trade and Industry and its Fisheries Agency are attempting to develop bioplastic substances that will decompose naturally in seawater.

Fisheries

All state waters (out to three nautical miles) are designated marine life conservation areas and subject to the Department of Land and Natural Resources' rules concerning the conservation of marine life. Commercial fishing requires a permit, and a special permit is needed for the Northwestern Hawaiian Islands, where rules can be adopted to ensure that fishery resources there will not be depleted. The state relies on the methods typically used by any fish and game agency to regulate the take of fish: gear regulation, open and closed seasons, catch and bag limits. In response to concerns about gillnets, the state

[9] 33 U.S.C. §§ 1901–12 (1988).

has banned their use in state waters and will confiscate such equipment. State statutes that regulate the take of wildlife, including the Hawaii Endangered Species Act, apply to wildlife in the ocean, but as a practical matter they are difficult to enforce at sea.

The Fishery Conservation and Management Act authorizes the federal government to regulate marine fisheries and establishes eight regional fishery councils to administer its programs.[10] The statute generally preserves state regulation in state waters and authorizes federal regulation out to 200 nautical miles. The Western Pacific Fisheries Management Council is made up of representatives of the State of Hawaii, American Samoa, Northern Marianas, Guam, and the National Marine Fisheries Service, and several appointed members who are usually associated with the commercial or recreational fishing industry. The council develops a management plan for each major fishery within its region which requires conservation and management. No plan can be approved without an environmental impact statement, a requirement that ensures input from agencies and the public. Plans for precious corals, lobsters, billfish, bottomfish, and seamount groundfish have been developed and approved by the federal Department of Commerce.

The Fishery Conservation and Management Act establishes national standards for fishery management plans. Fisheries must be managed to avoid long-term adverse effects on the marine environment and to avoid overfishing while achieving an "optimum yield." The councils are required to use the best scientific information available to establish their plans. Congress recognized that annual catches fluctuate widely, and the legislative history of the act indicates an intent to provide a margin of error as a buffer in favor of marine resource conservation. Though no court has ever directly addressed the subject, the national standards seem to require that fisheries be managed to provide sufficient food to maintain seabird populations.

The Fishery Conservation and Management Act gives the Western Pacific Fisheries Management Council exclusive management over fish and other marine life in the exclusive economic zone around Hawaii, but specifically exempts marine mammals, birds, and tunas. The exclusion of seabirds poses a problem. In the opinion of the solicitor of the Department of the Interior, the Migratory Bird Treaty Act cannot be enforced beyond the territorial sea. Because the status of the federal Endangered Species Act in the exclusive economic zone is similarly uncertain, Hawaiian seabirds seem to lack statutory protection throughout much of their feeding ranges. Even worse, the catch of skipjack tunas by foreigners is unregulated beyond the twelve-mile territorial sea. The United States asserts that its fishermen have the right to fish for tunas in the exclusive economic zones of South Pacific and Latin American nations because tunas are an international migratory resource and catches cannot be regulated by any single nation. A corollary of this policy allows foreigners to fish for tunas in the exclusive economic zone around Hawaii. The fishery

[10]16 U.S.C. §§ 1801–61 (1988).

management plan for billfish and associated species prohibits all drift gillnet fishing by foreign vessels within the exclusive economic zone.

Fishing in international waters beyond 200 nautical miles has been considered a freedom of the high seas for centuries. However, nations must comply with the environmental provisions of the Law of the Sea Treaty either as signatories or because those provisions have become customary international law. The Law of the Sea Treaty establishes a duty to protect and preserve the habitats of endangered species and requires fishing nations to consider the effects of fishing on species associated with or dependent on harvested species. As signatories to the treaty, Japan and the Republic of Korea should comply with its environmental impact statement and reporting requirements with respect to their high seas drift gillnet fisheries for squid. They should monitor and assess the effects of their fisheries on birds, mammals, and turtles caught in their gillnets and should make reports available to interested nations. The uncertainty of Taiwan's status as a nation increases the difficulties of forcing the Taiwanese to comply with customary international law.

Congress amended the Fishery Conservation and Management Act in 1987 to require the secretary of state to negotiate with nations that conduct drift gillnet fishing in international waters of the North Pacific. In mid-1989 the United States entered into agreements with Japan, Taiwan, and South Korea which allow U.S. biologists to assess the effects of such fisheries on U.S. marine resources, including seabirds. Japan has also agreed to reduce its gillnet fishery substantially. In December 1989 the United Nations General Assembly adopted a resolution calling for an end to drift gillnet fishing in international waters by mid-1992 unless fishing nations can prove the practice is not harmful.

Efficacy

The ocean is subject to little comprehensive regulation, management, or planning because it remains a relatively unexploited frontier. As most industrial uses of Hawaii's marine waters are speculative and may never be economically feasible, we have decades if not centuries to struggle with innovative means to protect the marine environment from industrial development. Most statutes that should protect marine resources are weakly enforced. The state is notorious for lax enforcement of environmental laws in its waters. Any efforts by the state to extend its jurisdiction from three to twelve nautical miles to conform with the federal expansion of the territorial sea must be accompanied by a major commitment of resources to monitor and protect those waters. Federal enforcement in the waters of the Northwestern Hawaiian Islands is limited to occasional Coast Guard patrols. Such surveillance no doubt deters some unauthorized activities, but the enforcement situation brings to mind the old Russian saying "Heaven is too high and the tsar is too far away."

Although the United States' tuna policy hamstrings the Western Pacific Fisheries Management Council, the council has made insufficient efforts to use the tools available to it to meet the conservation goals of the Fishery Conservation and Management Act. The council should develop a fishery management plan for tunas in the exclusive economic zone surrounding Hawaii to regulate U.S. fishermen. If fisheries for coastal pelagics, baitfish, or squid should be developed in the Northwestern Hawaiian Islands, their management plans should include mechanisms for emergency closure when substantial evidence indicates adverse effects on seabird colonies.

The council should improve its internal procedures, some of which might not withstand judicial review. It keeps few records, often does not identify the source of the information on which it bases its decisions, and relies on a scientific committee that includes members who lack scientific credentials. Some biologists refuse to participate in the council's activities because its decision making has appeared to be politicized if not arbitrary and capricious.

The state has insufficiently explored the use of marine sanctuaries as a means to manage and protect state waters. It has adopted a defensive state's-rights approach to this federal-state program, which it seems to view as an intrusion by the federal government. Instead, the state should set its own agenda and propose sanctuaries for which the Department of Land and Natural Resources could function as site manager. Marine sanctuaries in Hawaii could enhance resource management, protect seabirds and other wildlife, improve enforcement in protected waters, and garner federal funds for marine education and research. The waters offshore Kilauea Point and Crater Hill, Kauai, in addition to those of the Northwestern Hawaiian Islands, seem ideal for a marine sanctuary.

Federal enforcement of the Migratory Bird Act and Endangered Species Act in the exclusive economic zone requires express congressional authorization. Seabird management at sea is hampered by turf battles among federal agencies. The U.S. Fish and Wildlife Service theoretically has such authority, but as a terrestrial agency without oceangoing vessels it has minimal ability to conduct research and manage seabirds at sea. Amendment of the Fishery Conservation and Management Act to authorize the National Ocean and Atmospheric Administration to manage and conserve seabirds at sea, as it now does with regard to marine turtles, would be a great improvement.

Information is needed about the by-catch of seabirds, marine turtles, and marine mammals in the drift gillnet fisheries for squid in international waters. Fortunately, the U.S. State Department's negotiations with Japan, the Republic of Korea, and Taiwan were successful in 1989 and we will soon have the information necessary to assess these fisheries. Australia and the U.S. Congress are urging an international treaty to eliminate the use of high-seas drift nets throughout the world. Japan already prohibits such techniques in its own waters; we may hope that the international community will act before the devastation reaches a critical point. As S. Dillon Ripley of the Smithsonian Institution suggested a decade ago, an international commission on seabirds, similar to the International Whaling Commission, would be useful in efforts to deal with the problems of fisheries and seabirds on a global scale.

18 CONSERVATION DILEMMAS

The chief economist for one of the largest banks in Hawaii stood at the podium in an exclusive club in downtown Honolulu. A group of corporate attorneys had assembled for lunch to learn something about Hawaii's business climate. The economist began his address with the observation that Hawaii's economy is strikingly similar to most developing economies in the third world. It is built primarily on three pillars: tropical agriculture, military activities, and tourism. Of course, he added, unlike many developing nations, Hawaii does not have a corrupt government. His audience roared with laughter.

Residents of Hawaii are accustomed to reading newspaper accounts of corrupt practices: the U.S. Customs Service has seized $40,000 of undeclared jewelry from a former governor and close friend of Ferdinand Marcos; the uncovering of massive traffic-ticket fixing in the sheriff's office in Honolulu has implicated the present governor and the chief justice of the Hawaii Supreme Court; a federal audit of the Western Pacific Fisheries Management Council has found extensive financial irregularities. George Cooper and Gavan Daws's *Land and Power in Hawaii* documents the fact that public officials, including legislators, cabinet officers, judges, and labor leaders, have profited enormously from the development of Hawaii's land. Some of these people have rezoned property for their own financial benefit or that of their families and friends. The Bishop Estate, a perpetual educational trust and the largest private landowner in Hawaii, does not view the preservation of Hawaii's unique natural history as part of its public trust. Its five trustees award themselves annually almost $1 million *each* for nebulous services that include mustering a quorum of three to conduct official business once a week.

Hawaii's most lucrative crop is reputed to be *pakalolo* (marijuana), which is thought to yield more than pineapple and sugar cane combined and by some estimates earns between $3 billion and $10 billion a year. *Pakalolo* growers' crops encroach on conservation lands specifically designated as wildlife preserves, and innocent hikers sometimes encounter deadly booby traps, armed guards, or pit bulls trained to attack "intruders."

Even if Hawaii had an exemplary government, the elements of late-twentieth-century life which have destroyed much of the native Hawaiian ecosystems would continue to do so. Hawaii may always have the lion's share of America's endangered species. Charles Darwin observed long ago that island ecosystems are particularly sensitive to disturbance by alien species. Creatures that evolve in island environments usually experience less competition and predation than those that evolve in comparable continental habitats, and as a result introduced species are often competitively superior to insular ones. The success of introduced species on oceanic islands can be attributed in part to their ability to withstand disturbance by humans. Tropical seabirds and other Hawaiian wildlife evolved in the absence of terrestrial predators, and most cannot withstand the ravages of feral pigs, dogs, mongooses, and rats.

It is impossible to preserve every natural habitat in Hawaii, and, given its desirability as a place to live and visit, pressures on the land will increase. Hotels and resorts will be built whether or not fledgling shearwaters and petrels are drawn to their lights. The success or failure of human enterprise in Hawaii's seas will depend on economics, not on incidental harm to seabirds or other wildlife. In some sense, the problems seem to be intractable. Conservation problems that George C. Munro identified fifty years ago are similar to those of today.

The Role of Governments

Although federal legislation is needed to extend the authority of the Migratory Bird Treaty Act and the Endangered Species Act throughout the exclusive economic zone, current federal and state statutes generally provide ample statutory protection for Hawaii's wildlife. Most regulations are also satisfactory, but as the State of Hawaii has yet to codify its administrative rules, even attorneys have difficulty locating current environmental regulations. Most governmental conservation problems in Hawaii emanate from inadequate funding, poor implementation of policies, or weak enforcement of statutes. Americans are twice as likely to enjoy wildlife with binoculars or cameras as they are with rifles or fishing rods, and this comparison is further skewed in Hawaii, where each year the state hosts six tourists for each resident. Yet national studies indicate that $9 of every $10 spent on wildlife conservation in the United States primarily benefits animals that can be hunted or fished; nongame species such as seabirds and endangered species are left to shift for themselves in an era of tightening budgets.

Hawaii's legislature stated laudable goals to protect and preserve the wildlife of Hawaii in its state plan, but they are drowned in a nonprioritized laundry list of other economic and social goals. The legislature has repeatedly refused to allow voluntary checkoffs on state income tax returns to fund nongame wildlife programs. It has repeatedly declined to require the Department of Land and Natural Resources to provide detailed audits of state wildlife programs to determine whether federal matching funds have been properly spent, despite considerable evidence of mismanagement or misappropriation. The Wildlife Management Institute, a national organization of professional wildlife managers, would gladly lend its expertise to the task of evaluating and improving Hawaii's wildlife programs, as it has done for forty other states. The state refuses to issue an invitation. The wildlife programs of the Department of Land and Natural Resources are poorly funded and understaffed. Morale is so low that its best wildlife biologists and managers suffer through only a few years of the stifling bureaucracy created by the state's aparatchiks before transferring to other positions, usually outside state government. The state's financial support for professional expertise is so low that the American Fishery Society ranks Hawaii last among the fifty states in respect to the salaries it pays its fishery biologists. State fishery biologists fare better in such landlocked states as Oklahoma, New Mexico, and North Dakota than in Hawaii.

Federal funding of wildlife programs, while superior to the state's, remains inadequate. Federal support is essential to Hawaii. The state has over one-quarter of this nation's endangered species, and it is simply impossible for Hawaii's one million residents to shoulder such a disproportionate burden of a national problem. Congress enacted the Fish and Wildlife Conservation Act in 1980 as a matching-grant program for nongame conservation but has yet to fund it, rendering its passage a hollow gesture.[1] The U.S. Fish and Wildlife Service in Hawaii has never acquired a vessel to study and manage seabirds at sea or even to enforce its own regulations at its Northwestern Hawaiian Islands wildlife refuge. The agency has never established a wildlife cooperative research unit at the University of Hawaii to provide long-term technical support to both the state and federal governments, and its research division's work in Hawaii has declined precipitously since the early 1980s.

A major problem for the state government is that its efforts to manage its natural resources are dwarfed by the politics of land and power in the Department of Land and Natural Resources. The state government systematically excludes individuals with professional qualifications in natural resource management from high-level decision making. Biologists and managers employed by the Department of Land and Natural Resources exert some influence on policies promulgated by the board, but their technical expertise is questionable and their influence minimal. After almost three decades of statehood, no governor has appointed a single wildlife manager, biologist, fishery manager, or

[1] 16 U.S.C. §§ 2901–12 (1988).

forester to the six-member Board of Land and Natural Resources. Functioning as an executive board, it is charged with the management and administration of public lands and wildlife resources and makes virtually all significant decisions concerning the use of conservation land. It establishes the policies and priorities of the Department of Land and Natural Resources. The nine-member Land Use Commission has never included a person with expertise in natural resource management, although its implemention of the statewide zoning statute and its decisions concerning the boundaries of conservation districts would seem to require such individuals. The Land Use Commission and the Board of Land and Natural Resources are like a board of agriculture with no agronomist or a board of education that lacks a single member with professional qualifications in education.

Realtors, developers, security guards, and the International Longshoremen's and Warehousemen's Union have been well represented on both the Board of Land and Natural Resources and the Land Use Commission. When individuals with skills and expertise in natural resource management are systematically excluded, poor or uninformed decisions are assured. The role of the International Longshoremen's and Warehousemen's Union is especially peculiar; Democratic governors during the past two decades have ensured that the voice of this powerful labor union is heard when natural resource decisions are to be made. Members of equivalent state boards in ten states are nominated by citizen groups. Michigan's commissioners who oversee its Department of Natural Resources are required by statute to have professional qualifications in some aspect of natural resource management. Hawaii's legislators and officials, however, seem to believe that the Department of Land and Natural Resources and the Land Use Commission should be directed by generalists rather than experts. Although none has ever expressly said so, governors may intentionally exclude individuals with knowledge of natural resource management from consideration for appointment. It seems inescapable that the agenda has been to minimize professional management, thereby maximizing political influence over natural resource decision making.

The Agencies

State and federal agencies have enormous influence over the implementation of wildlife laws. Formally through regulation and informally in day-to-day decisions they establish policies in the large interstices among black-letter statutes. Lawsuits that challenge agency decisions always face an uphill battle because courts tend to defer to the agencies in the areas of their presumed expertise. Unfortunately, senior officials in many agencies advance in their organizations more by virtue of political connections than through knowledge and skill.

The U.S. Fish and Wildlife Service is administered largely by individuals who are oriented to the service's historical programs, which focus on the consumptive uses of wildlife. The inordinate influence that hunting and fishing organizations exert over its policies and budget priorities stems largely from the fact that senior officials usually have been selected for their backgrounds in the management of game rather than of the ecosystem. Individuals whose careers have focused on waterfowl production rather than the preservation of endangered species naturally view hunters and fishermen as their primary clientele. Advancement to senior levels in the U.S. Fish and Wildlife Service tends to use good-ol'-boy procedures that reinforce a bias toward achieving full bags for hunters to the exclusion of other wildlife programs. Within the agency it is common knowledge that many "open" positions listed in the green sheet, the internal employment newsletter, have been described in such a way as to fit the qualifications and experience of preselected individuals. An inevitable result of such a system is that the problems of Hawaii's seabirds and endangered species are foreign to senior U.S. Fish and Wildlife Service officials in Washington and in the Portland regional office. These people tend to lack both the experience and the inclination to work on such problems.

The employees of the U.S. Fish and Wildlife Service and the Department of Land and Natural Resources are an inbred lot. The agencies should cast wider nets when they recruit senior administrators, most of whom attain their positions by rising through the ranks of the agency without obtaining the benefits of the cross-fertilization that occurs when a career encompasses a variety of organizations. This problem is exacerbated in the Department of Land and Natural Resources by its small size and the limited opportunities it offers for professional growth. Many senior managers of the U.S. Fish and Wildlife Service and the Department of Land and Natural Resources are not members of any of the societies in the fields in which they should be professionals: ecology, wildlife management, mammalogy, ornithology, botany. Few possess significant scientific credentials or have names that would be recognized by biologists in these fields. Senior U.S. Fish and Wildlife Service administrators in Washington can seem to be indistinguishable from their counterparts in agencies that deal with veterans' affairs, government procurement, or social security benefits.

Top-drawer managers with extensive experience as both scientists and administrators exist and, if recruited, would gladly assume responsible positions with the U.S. Fish and Wildlife Service or the Department of Land and Natural Resources. Deans of colleges with strong programs in natural resource management, executives in private conservation organizations, senior congressional staff, and officials of other state and federal natural resource agencies are systematically excluded from senior management positions. Unlike other federal scientific and technocratic agencies, the U.S. Fish and Wildlife Service rarely if ever recruits top management from outside the agency. The Department of Land and Natural Resources is so hamstrung by state civil service procedures that senior positions are offered to individuals with minimal qualifications merely because they are already state employees. Neither agency advertises employment opportunities in *Science*, the weekly magazine of the American

Association for the Advancement of Science, which is universally used as a job vacancy bulletin board by scientific and technical organizations, including state and federal natural resource agancies. If all positions ranked GS-14 and higher in the U.S. Fish and Wildlife Service and all positions of bureau chief and higher in the Department of Land and Natural Resources were widely advertised and a third were filled by well-qualified individuals from outside the agency, the increase in professionalism and improvement in programs could be immediate and dramatic.

Paradoxically, neither the Department of Land and Natural Resources nor the U.S. Fish and Wildlife Service seems to appreciate the need for scientists. The state wildlife plan gives a low priority to research and the Department of Land and Natural Resources looks to the University of Hawaii, the U.S. Fish and Wildlife Service, and the National Marine Fisheries Service for technical skills and support. The sound, statistically defensible biological information demanded by scientists is too often resented by managers, who prefer instead quick-and-dirty "research" to solve immediate problems. Whether solutions based on such studies are efficacious is seldom asked. U.S. Fish and Wildlife Service biologists have difficulty in traveling to important conferences and symposia, whereas agency funds routinely enable senior bureaucrats to travel to distant meetings on matters that could be handled with a ten-minute conference call. Biologists in the Department of Land and Natural Resources are virtually barred from attending scientific meetings outside of Hawaii, with the result that their ability to remain abreast of professional developments is severely hampered.

Why do natural resource agencies seem to have so little appreciation for research applications and scientific solutions to their problems? One reason may be the increasing politicization of natural resource decisions. Objective scientific facts can be insurmountable barriers to political maneuvering and off-the-record arrangements. Another may be that scientists tend to owe greater allegiance to science and the "search for truth" than to an employer. Development of new techniques and methods of wildlife management are being deemphasized at a time when they are needed to mitigate the multiple onslaughts against Hawaiian wildlife and its habitats. Many ecosystem management problems will yield to technical solutions. When agencies avoid developing new technologies, developers, conservationists, and natural resource managers are left to fight the same old environmental battles that might be avoided altogether.

Techniques could be developed to manipulate a seabird colony so that it could be relocated when a breeding site must give way to another use. Genetic engineers are working on a blue-green algae that kills mosquito larvae, which could provide a means to control mosquitos and increase lowland habitat for birds that have been ravaged by avian malaria. The eradication of rats from even small islands has proved to be a formidable task, and some populations have developed resistance to anticoagulant rodenticides. The largest island from which rats have been eradicated is Otata, a 22-hectare island in the Noises group, New Zealand. The islets offshore north Oahu are excellent laboratories for improvement of techniques to remove introduced predators. All too soon we may need those methods on Laysan or Lisianski. Wildlife techniques developed in Hawaii could be used in the developing economies of Asia, Latin America, and Africa, where tropical nations desperately need technical assistance to manage their natural resources.

Private Organizations

A century and a half ago the French writer Alexis de Tocqueville marveled at the grass-roots organizations that sprang forth in the United States to solve problems. Such groups continue to proliferate today. Private conservation organizations can solve public problems creatively, avoiding the rigidity and delays of governmental agencies. They can be especially effective when they actively oversee, advise, and cooperate with natural resource agencies. Hawaii would be an excellent location in which to experiment with the privatization of parks and wildlife refuges. Conservation organizations could directly manage state and federal lands under suitable long-term leases. Successes at Manana Island or Kawainui Marsh could revolutionize the management of public lands.

A fundamental problem for wildlife conservation in Hawaii has been the relative ineffectiveness of Hawaii's plethora of private conservation organizations, which for decades were poorly organized and relied almost exclusively on volunteer efforts. Cooper and Daws summarize the situation since statehood:

> In general conservationist terms, there were throughout the Democratic years a number of Honolulu-based respectable middle-class groups which conscientiously expressed anti-development opinions at public hearings, but which by themselves had no clearly large-scale influence: the Conservation Council, the Audubon Society, the Outdoor Circle, and so on.

The remnants of Hawaii's natural history are national and international resources, and their protection deserves and requires paid professionals, including natural resource administrators, biologists, and fund raisers dedicated to planning and implementing long-term conservation strategies.

The establishment of the Hawaii office of The Nature Conservancy in the early 1980s represented the first recognition by a national conservation organization of the problems and challenges that Hawaii faces. The Nature Conservancy has been enormously successful in motivating community leaders in both private and public sectors to promote land acquisition and to enhance appropriations for the state's natural area reserves program. Its heritage program has established a biological data base to help answer questions on natural resources. Despite its obvious successes, The Nature Conservancy has institutional limitations. Its ability to work with large corporations could be seriously undermined if it became embroiled in public controversies. Consequently, The Nature Conservancy cannot effectively oversee recalcitrant government agencies or such landowners as the Bishop Estate.

Nineteen-eighty-eight was a watershed year for the establishment of private conservation organizations in Hawaii. The Trust for Public Land closed its first transaction in Hawaii, successfully brokering the acquisition of seabird colonies at Crater Hill and Mokolea Point by the Kilauea Point National Wildlife Refuge. Grants from the MacArthur Foundation enabled the National Audubon Society, the Natural Resources Defense Council, and the Sierra Club Legal Defense Fund to open offices in Honolulu. The entrée of sophisticated national environmental organizations is an extremely encouraging development that over time should dramatically improve the conservation landscape.

Environmental organizations in Hawaii tend to be constituted largely of individuals who have moved to the islands from elsewhere. It seems inescapable that conservation values are poorly developed in individuals born and raised in Hawaii, with the exception of those who have spent a significant portion of their lives elsewhere. While education is ultimately the remedy for this situation, the task encompasses far more than improving lessons for schoolchildren. The challenge of educating adults is exacerbated by adult illiteracy rates in the rural islands (all islands except Oahu), which exceed one-third. Among fishermen and farmers, the people most likely to encounter wildlife in their daily routines, functional illiteracy approaches one-half. Motivating illiterates to protect Hawaii's natural history is an especially difficult task. All too often such people retain psychological baggage from a plantation society that disappeared decades ago, and resent conservationists as intrusive mainlanders. The islands' demographics, however, are changing rapidly: by 1980 almost half of Hawaii's residents had been born beyond the state's boundaries. Many of the millions of people who visit Hawaii each year are interested in its environmental problems and are potential allies on conservation issues if they can be informed and motivated. Hawaiian conservation organizations may find that nationwide fund-raising efforts for Hawaiian projects are especially successful.

If private organizations are to influence government effectively, they must be able to bring citizens' suits to enforce federal and state statutes. The state courts of Hawaii have countered the reluctance of state agencies to enforce their own laws by providing broad standing to organizations and individuals to challenge state environmental policies and actions. Many federal environmental statutes specifically provide for a private right of action, allowing concerned organizations and individuals to bring suits to enforce compliance. Lawsuits are expensive, especially in state courts, which cannot order the state or a corporation to pay a successful litigant's attorneys' fees and costs. Private organizations that bring such suits usually lack the resources to pay attorneys' fees and must rely on pro bono publico representation. Such federal statutes as the Endangered Species Act, by contrast, allow an environmental organization to recoup its attorneys' fees. As a consequence, attorneys who file a suit concerning a violation of an endangered species act always prefer to sue in federal rather than state court.

The Challenge

Most large, conspicuous seabird colonies in Hawaii have been acquired by governmental agencies or private conservation organizations. The protection of breeding areas for species whose nest sites or habitats are less conspicuous has been largely accidental. Although it is stylish to believe that all is well once habitat has been acquired, such assumptions are rarely justified, especially in Hawaii. Many areas need active, even intensive management to prevent habitat destruction, which can be irreversible. Funds to manage and improve management techniques are crucial.

The challenge in Hawaii is to base environmental decisions that affect wildlife on good scientific information. The Board of Land and Natural Resources and the public are often ignorant of the facts on which such decisions should be based. Too often battles have been fought on the basis more of emotion than of information, with the resulting decisions based on brute strength. A National Research Council report on prediction of environmental effects found that too many arguments about development projects are based on prejudice rather than knowledge.

> There are a lot of people on both sides of the arguments who find it convenient not to know what actually happens when a project goes ahead. That way they can go into the next environmental battle with their positions unchanged.

Ecological science and its practical applications would benefit if projects were treated as experiments. Careful monitoring is essential if we are to understand the effects of a project, test predictions made in an environmental impact statement, measure changes in baseline conditions, and detect cumulative effects. Government resources are increasingly limited, and effective conservation efforts require private organizations to assume greater roles in raising funds, developing techniques, assembling information, and presenting it to decision makers.

Ultimately decisions concerning the management of Hawaiian wildlife are political. Many conservationists espouse egalitarian views and seemingly believe that all species should be protected equally, but such an approach spreads conservation resources thin and can bring poor results. The public does not believe that all species are equal. Individuals have different value systems and draw their own lines, but in the continuum of life forms from mammals, birds, and fish through plants, insects, flatworms, fungi, amoebas, bacteria, viruses, and polypeptides virtually everyone would choose to protect some natural forms to the exclusion of others. The public would have no difficulty in choosing to preserve Bengal tigers rather than pupfish if such a choice were necessary. It may not seem "fair," but a Gallup poll would undoubtedly find more people willing to spend money to save a Laysan albatross colony than a sooty storm-petrel colony. Increased public education would help many people to make such decisions, but time, energy, and dollars are finite and not all battles can be fought, let alone won.

APPENDIX

Some Common and Scientific Names

Plants

beach heliotrope	*Messerschmidia [Tournefortia] argentea*
beach magnolia	*Scaevola taccada*
beach morning-glory	*Ipomoea* spp.
bunchgrass	*Eragrostis* spp.
carpetweed	*Sesuvium portulacastrum*
golden crown-beard	*Verbesina enceloides*
goosefoot	Chenopodiacae
ilima	*Sida fallax*
ironwood	*Casuarina equisetifolia*
loulou fan palm	*Pritchardia remota*
mesquite	*Prosopis* spp.
nightshade	*Solanum* spp.
ohai	*Sesbania tomentosa*
puncture vine	*Tribulus cistoides*
purslane	*Portulaca* spp.
sandalwood	*Santulum* spp.
San Francisco grass	*Ammophila arenaria*
uluhe fern	*Dicranopteris linearis*
wild mustard	*Brassica campestris*

Birds

American golden plover	*Pluvialis dominica*
black-crowned night heron	*Nycticorax nycticorax*
black-footed albatross	*Diomedea nigripes*

black noddy	*Anous minutus [tenuirostris]*
blue-gray noddy	*Procelsterna cerulea*
Bonin petrel	*Pterodroma hypoleuca*
bristle-thighed curlew	*Numenius tahitiensis*
brown booby	*Sula leucogaster*
brown noddy	*Anous stolidus*
bulbul	*Pycnonotus* spp.
Bulwer's petrel	*Bulweria bulwerii*
cattle egret	*Bubulcus ibis*
Christmas shearwater	*Puffinus nativitatis*
common mynah	*Acridotheres tristis*
dark-rumped petrel	*Pterodroma phaeopygia sandwichensis*
gray-backed tern	*Sterna lunata*
great frigatebird	*Fregata minor*
Harcourt's (band-rumped) storm-petrel	*Oceanodroma castro*
Hawaiian stilt	*Himantopus himantopus knudseni*
herring gull	*Larus argentatus*
Laysan albatross	*Diomedea immutabilis*
Laysan finch	*Telespiza cantans*
lesser frigatebird	*Fregata ariel*
magenta petrel	*Pterodroma magentae*
magnificent frigatebird	*Fregata magnificens*
masked booby	*Sula dactylatra*
Newell's shearwater	*Puffinus [auricularis] newelli*
Nihoa finch	*Telespiza ultima*
Nihoa millerbird	*Acrocephalus familiaris kingi*
northern fulmar	*Fulmarus glacialis*
red-billed tropicbird	*Phaethon aethereus*
red-footed booby	*Sula sula*
red-tailed tropicbird	*Phaethon rubricauda*
rock dove	*Columba livia*
ruddy turnstone	*Arenaria interpres*
sanderling	*Calidris alba*
short-tailed albatross	*Diomedea albatrus*
sooty storm-petrel	*Oceanodroma tristrami*
sooty tern	*Sterna fuscata*
tubenose	Procellariiformes
wandering tattler	*Heteroscelus incanus*
wedge-tailed shearwater	*Puffinus pacificus*
white-tailed tropicbird	*Phaethon lepturus*
white tern	*Gygis alba*
white-throated storm-petrel	*Nesofregetta fuliginosa*

Mammals

black rat	*Rattus rattus*
common dolphin	*Delphinus delphis*

European rabbit	*Oryctolagus cuniculus*
Hawaiian monk seal	*Monachus schauinslandi*
mongoose	*Herpestes auropunctatus*
mouse	*Mus musculus*
Norway rat	*Rattus norvegicus*
Polynesian rat	*Rattus exulans*
spinner dolphin	*Stenella longirostris*
spotter dolphin	*Stenella attenuata*

Fish

anchovy	*Stolephorus buccaneeri*
balloonfish	*Lagocephalus lagocephalus*
blue marlin	*Makaira nigricans*
bristlemouth	*Gonorhynchus gonorhynchus*
Cuvier's flyingfish	*Cypselurus speculiger*
dolphinfish	*Coryphanea* spp.
fantail filefish	*Pervagor spilosoma*
five-horned cowfish	*Lactoria fornasini*
flyingfish	Exocoetidae
flying gurnard	*Dactyloptena orientalis*
Forster's lizardfish	*Trachinocephalus myops*
goatfish	Mullidae
goby	*Ptereleotris heteropterus*
green halfbeak	*Euleptorhamphus viridis*
Gregory's fish	*Gregoryina gygis (Cheilodactylus vittaus)*
halfbeak	Hemiramphidae
hatchetfish	Sternoptychidae
herring	*Spratelloides delicatulus*
jack	*Caranx* spp.
lanternfish	Myctophidae
Linne's flyingfish	*Exocoetus volitans*
little tuna	*Euthynnus affinis*
lizardfish	Synodontidae
mackerel scad	*Decapterus* spp.
needlefish	Belonidae
Pacific saury	*Cololabis saira*
round herring	*Spratelloides delicatulus*
rudderfish	*Kyphosus bigibbus*
saury	*Cololabis* spp.
short-winged flyingfish	*Parexocoetus brachypterus*
skipjack tuna	*Katsuwonus pelamis*
snake mackerel	*Gempylus serpens*
squirrelfish	Holocentridae
striped hawkfish	*Cheilodactylus vittatus*
swordfish	Xiphiidae

truncated sunfish	*Ranzania laevis*
yellowfin tuna	*Thunnus albacares*

Invertebrates

flying squid	Ommastrephidae (esp. *Stenoteuthis* [*Symplectoteuthis*] *oualaniensis*)
krill	*Euphausia superba*
sea strider	*Halobates sericeus*
wind sailor	*Velella velella*

SELECTED BIBLIOGRAPHY

General

Amerson, A. Binion, Jr. 1971. *The Natural History of French Frigate Shoals, Northwestern Hawaiian Islands.* Atoll Research Bulletin no. 150. Washington, D.C.

——, Roger B. Clapp, and William O. Wirtz II. 1974. *The Natural History of Pearl and Hermes Reef, Northwestern Hawaiian Islands.* Atoll Research Bulletin no. 174. Washington, D.C.

Bean, Michael J. 1983. *The Evolution of National Wildlife Law.* New York.

Clapp, Roger B., and Eugene Kridler. 1977. *The Natural History of Necker Island, Northwestern Hawaiian Islands.* Atoll Research Bulletin no. 206. Washington, D.C.

——, ——, and Robert R. Fleet. 1977. *The Natural History of Nihoa Island, Northwestern Hawaiian Islands.* Atoll Research Bulletin no. 207. Washington, D.C.

—— and William O. Wirtz II. 1975. *The Natural History of Lisianski Island, Northwestern Hawaiian Islands.* Atoll Research Bulletin no. 186. Washington, D.C.

Dole, Sanford B. 1869. A Synopsis of the Birds Hitherto Described From the Hawaiian Islands. *Proceedings of the Boston Society of Natural History* 12:294–309.

Ely, Charles A., and Roger B. Clapp. 1973. *The Natural History of Laysan Island, Northwestern Hawaiian Islands.* Atoll Research Bulletin no. 171. Washington, D.C.

Fisher, Walter K. 1906. *Birds of Laysan and the Leeward Islands, Hawaiian Group.* Bulletin of the U.S. Fish Commission, vol. 23. Washington, D.C.

Harrison, Peter. 1983. *Seabirds: An Identification Guide.* Boston.

Hawaii Audubon Society. 1940–1989. *'Elepaio.* Honolulu.

King, Warren B. 1967. *Preliminary Smithsonian Identification Manual: Seabirds of the Tropical Pacific Ocean.* Washington, D.C.

Munro, George C. 1960. *Birds of Hawaii.* Rutland, Vt.

Murphy, Robert Cushman. 1936. *Oceanic Birds of South America.* New York.

Woodward, Paul W. 1972. *The Natural History of Kure Atoll, Northwestern Hawaiian Islands.* Atoll Research Bulletin no. 164. Washington, D.C.

1. The Islands

Armstrong, R. Warwick, ed. 1983. *Atlas of Hawaii*. Honolulu.
Carlquist, Sherwin. 1980. *Hawaii: A Natural History*. Honolulu.
Couper, Alastair D. 1983. *Atlas of the Oceans*. London.
Grigg, Richard W. 1988. Paleoceanography of coral reefs in the Hawaiian-Emperor chain. *Science* 240:1737–44.
Olson, Storrs L., and Helen F. James. 1982. *Prodromus of the Fossil Avifauna of the Hawaiian Islands*. Smithsonian Contributions to Zoology no. 365. Washington, D.C.
Rothschild, Walter. 1893–1900. *The Avifauna of Laysan and the Neighboring Islands*. London.
Wetmore, Alexander. 1925. Bird life among lava rock and coral sand. *National Geographic Magazine* 48, no. 7:76–108.

2. The Sea

Armstrong, R. Warwick, ed. 1983. *Atlas of Hawaii*. Honolulu.
Beebe, William. 1926. *The Arcturus Adventure*. New York.
Couper, Alastair D. 1983. *Atlas of the Oceans*. London.
Harrison, Craig S., Thomas S. Hida, and Michael P. Seki. 1983. *Hawaiian Seabird Feeding Ecology*. Wildlife Monograph no. 85. Bethesda, Md.
Hirota, J., S. Taguchi, R. F. Shuman, and A. E. Jahn. 1980. Distributions of plankton stocks, productivity, and potential fishery yield in Hawaiian waters. In *Proceedings of the Symposium on Status of Resource Investigations in the Northwestern Hawaiian Islands*, ed. Richard W. Grigg and Rose T. Pfund. Honolulu.
Uchida, Richard N., and James H. Uchiyama, eds. 1986. *Fishery Atlas of the Northwestern Hawaiian Islands*. NOAA Technical Report NMFS 38. Honolulu.

3. The Humans

Armstrong, R. Warwick, ed. 1983. *Atlas of Hawaii*. Honolulu.
Bryan, E. H., Jr. 1980. *The Northwestern Hawaiian Islands: An Annotated Bibliography*. Honolulu: U.S. Fish and Wildlife Service.
Olson, Storrs L., and Helen F. James. 1982. *Prodromus of the Fossil Avifauna of the Hawaiian Islands*. Smithsonian Contributions to Zoology no. 365. Washington, D.C.

4. Origin and Adaptations of Hawaiian Seabirds

Harrison, C. J. O. 1978. *Bird Families of the World*. New York.
Lockley, R. M. 1974. *Ocean Wanderers*. Harrisburg, Pa.
Nelson, Bryan. 1979. *Seabirds: Their Biology and Ecology*. New York.

5. Populations

Ashmole, N. P. 1963. The regulation of numbers of tropical oceanic birds. *Ibis* 103b:458–73.
Bartsch, Paul. 1922. A visit to Midway Island. *Auk* 39:481–88.
Diamond, A. W. 1978. Feeding strategies and population size in tropical seabirds. *American Naturalist* 112:215–23.
Harrison, Craig S., Maura B. Naughton, and Stewart I. Fefer. 1984. The status and conservation of seabirds in the Hawaiian archipelago and Johnston Atoll. In *Status and Conservation of*

the World's Seabirds, ed. J. P. Croxall, P. G. H. Evans, and R. W. Schreiber. ICBP Technical Publication no. 2. Cambridge, U.K.
Nelson, Bryan. 1979. *Seabirds: Their Biology and Ecology*. New York.

6. *Breeding Ecology*

Nelson, Bryan. 1979. *Seabirds: Their Biology and Ecology*. New York.
Nelson, J. B. 1984. Contrasts in breeding strategies between some tropical and temperate marine Pelecaniformes. In *Tropical Seabird Biology*, ed. Ralph W. Schreiber. Studies In Avian Biology no. 8. Los Angeles.
Whittow, G. C. 1984. Physiological ecology of incubation in tropical seabirds. In *Tropical Seabird Biology*, ed. Ralph W. Schreiber. Studies in Avian Biology no. 8. Los Angeles.

7. *Feeding Ecology*

Au, David W. K., and Robert L. Pitman. 1986. Seabird interactions with dolphins and tuna in the Eastern Tropical Pacific. *Condor* 88:304–17.
Beebe, William. 1924. *Galápagos: World's End*. New York.
Harrison, Craig S., Thomas S. Hida, and Michael P. Seki. 1983. *Hawaiian Seabird Feeding Ecology*. Wildlife Monograph 85. Bethesda, Md.
—— and Michael P. Seki. 1987. Trophic relationships among tropical seabirds at the Hawaiian Islands. In *Seabirds: Feeding Ecology and Role in Marine Ecosystems*, ed. J. P. Croxall. Cambridge, U.K.
Seki, Michael P., and Craig S. Harrison. 1989. Feeding ecology of two subtropical seabird species at French Frigate Shoals, Hawaii. *Bulletin of Marine Science* 45:52–67.

8. *Pelagic Ecology: Life at Sea*

Gould, Patrick Jerry. 1971. Interactions of seabirds over the open ocean. Ph.D. dissertation, University of Arizona, Tucson.
King, Warren B. 1970. *The Trade Wind Zone Oceanography Pilot Study*, pt. VII: *Observations of Seabirds, March 1964 to June 1965*. Special Scientific Report—Fisheries no. 586, U.S. Fish and Wildlife Service. Washington, D.C.
——, ed. 1974. *Pelagic studies of seabirds in the Central and Eastern Pacific Ocean*. Smithsonian Contributions to Zoology no. 158. Washington, D.C.
Pitman, Robert L. 1986. *Atlas of Seabird Distribution and Relative Abundance in the Eastern Tropical Pacific*. NOAA-NMFS-Southwest Fisheries Center Administrative Report LJ-86–02C. La Jolla, Calif.
Rothschild, Walter. 1893–1900. *The Avifauna of Laysan and the Neighboring Islands*. London.

9. *Albatrosses: Family Diomedeidae*

Fisher, Harvey I. 1971. The Laysan albatross: Its incubation, hatching, and associated behaviors. *Living Bird* 10:19–78.
——. 1975. Mortality and survival in the Laysan albatross, *Diomedea immutabilis*. *Pacific Science* 29:279–300.
—— and Mildred L. Fisher. 1969. The visits of Laysan albatrosses to the breeding colony. *Micronesica* 5:173–221.
Hasegawa, Hiroshi. 1984. Status and conservation of seabirds in Japan, with special attention to the short-tailed albatross. In *Status and Conservation of the World's Seabirds*, ed. J. P.

Croxall, P. G. H. Evans, and R. W. Schreiber. ICBP Technical Publication no. 2. Cambridge, U.K.

Robbins, Chandler S., and Dale W. Rice. 1974. Recoveries of banded Laysan albatrosses (*Diomedea immutabilis*) and black-footed albatrosses (*D. nigripes*). In *Pelagic Studies of Seabirds in the Central and Eastern Pacific Ocean*, ed. Warren B. King. Smithsonian Contributions to Zoology no. 158. Washington, D.C.

10. *Shearwaters and Gadfly Petrels: Family Procellariidae*

Grant, Gilbert S., John Warham, Ted N. Pettit, and G. Causey Whittow. 1983. Reproductive behavior and vocalizations of the Bonin petrel. *Wilson Bulletin* 95:522–39.

Imber, M. J. 1985. Origins, phylogeny, and taxonomy of the gadfly petrels *Pterodroma* spp. *Ibis* 127:197–229.

Palmer, Ralph S. 1962. *Handbook of North American Birds*, vol. 1. New Haven, Conn.

Rothschild, Walter. 1893–1900. *The Avifauna of Laysan and the Neighboring Islands*. London.

Sievert, Paul R., Louis Sileo, and Stewart I. Fefer. In press. Prevalence and characteristics of plastic ingested by Hawaiian seabirds. In *Proceedings of the Second International Conference on Marine Debris*, ed. Richard S. Shomura. Honolulu.

Simons, Theodore R. 1985. Biology and behavior of the endangered Hawaiian dark-rumped petrel. *Condor* 87:229–45.

Sincock, John L., and Gerald E. Swedberg. 1969. Rediscovery of the nesting grounds of Newell's Manx shearwater (*Puffinus puffinus newelli*), with initial observations. *Condor* 71:69–71.

11. *Storm-Petrels: Family Oceanitidae*

Allan, R. G. 1962. The Madeiran storm-petrel *Oceanodroma castro*. *Ibis* 103b:274–95.

Harris, M. P. 1969. The biology of storm-petrels in the Galapagos Islands. *Proceedings of the California Academy of Science* (Series 4) 37:95–166.

Harrison, Craig S., Thomas C. Telfer, and John L. Sincock. 1990. The status of Harcourt's storm-petrel (*Oceanodroma castro cryptoleucura*) in Hawaii. *'Elepaio* 50.

Naveen, Ron. 1981–82. Storm-petrels of the world. *Birding* 13 and 14.

Rauzon, Mark J., Craig S. Harrison, and Sheila Conant. 1985. The status of the sooty storm-petrel in Hawaii. *Wilson Bulletin* 97:390–92.

12. *Frigatebirds: Family Fregatidae*

Hadden, F. C. 1941. Midway Islands. *Hawaiian Planters' Record* 45:179–221.

Nelson, J. B. 1984. Contrasts in breeding strategies between some tropical and temperate marine Pelecaniformes. In *Tropical Seabird Biology*, ed. Ralph W. Schreiber. Studies in Avian Biology no. 8. Los Angeles.

Nelson, J. Bryan. 1975. The breeding biology of frigatebirds—a comparative review. *Living Bird* 14:113–55.

13. *Boobies: Family Sulidae*

Dorward, D. F. 1962. Behavior of boobies, *Sula* spp. *Ibis* 103b:221–34.

——. 1962. Comparative behavior of the white booby and the brown booby, *Sula* spp., at Ascension. *Ibis* 103b:174–220.

Kepler, Cameron B. 1969. Breeding biology of the blue-faced booby (*Sula dactylatra personata*) on Green Island, Kure Atoll. Nuttall Ornithology Club Publication no. 8. Cambridge, Mass.

Nelson, J. B. 1984. Contrasts in breeding strategies between some tropical and temperate marine Pelecaniformes. In *Tropical Seabird Biology*, ed. Ralph W. Schreiber. Studies in Avian Biology no. 8. Los Angeles.
Nelson, J. Bryan. 1978. *The Sulidae: Gannets and Boobies*. Oxford.
Palmer, Ralph S. ed. 1962. *Handbook of North American Birds*, vol. 1. New Haven, Conn.

14. Tropicbirds: Family Phaethontidae

Fleet, Robert R. 1974. *The Red-Tailed Tropicbird on Kure Atoll*. Ornithological Monograph 16. Washington, D.C.
Howell, Thomas R. 1978. *Ecology and Reproductive Behavior of the Gray Gull of Chile and of the Red-Tailed Tropicbird and White Tern of Midway Island*. National Geographic Society Research Reports. Washington, D.C.
Nelson, J. B. 1984. Contrasts in breeding strategies between some tropical and temperate marine Pelecaniformes. In *Tropical Seabird Biology*, ed. Ralph W. Schreiber. Studies in Avian Biology no. 8. Los Angeles.
Stonehouse, Bernard. 1963. The tropic birds (Genus *Phaethon*) of Ascension Island. *Ibis* 103b:124–61.

15. Terns and Noddies: Family Laridae (Subfamily Sterninae)

Cullen, J. M., and N. P. Ashmole. 1963. The black noddy *Anous tenuirostris* on Ascension Island, pt. 2, Behaviour. *Ibis* 103b:423–46.
Dinsmore, James J. 1971. Sooty tern behavior. *Bulletin of the Florida State Museum* 16:129–79.
Gould, Patrick J. 1974. Sooty tern (*Sterna fuscata*). In *Pelagic Studies of Seabirds in the Central and Eastern Pacific Ocean*, ed. Warren B. King. Smithsonian Contributions to Zoology no. 158. Washington, D.C.
Howell, Thomas R. 1978. *Ecology and Reproductive Behavior of the Gray Gull of Chile and of the Red-Tailed Tropicbird and White Tern of Midway Island*. National Geographic Society Research Reports. Washington, D.C.
Miles, Dorothy H. 1986. White terns breeding on Oahu, Hawaii. *'Elepaio* 46:171–75.
Rauzon, Mark J., Craig S. Harrison, and Roger B. Clapp. 1984. Breeding biology of the blue-gray noddy. *Journal of Field Ornithology* 55:309–21.
Tinker, Spencer Wilkie. 1978. *Fishes of Hawaii*. Honolulu.
Watson, John B. 1908. The behavior of noddy and sooty terns. *Papers from the Tortugas Laboratory of the Carnegie Institute* 2:187–255. Washington, D.C.

16. Conservation on the Islands

Atkinson, I. A. E. 1985. The spread of commensal species of *Rattus* to oceanic islands and their effects on island avifaunas. In *Conservation of Island Birds*, ed. P. J. Moors. ICBP Technical Publication no. 3. Cambridge, U.K.
Callies, David. 1985. *Regulating Paradise: Land Use Controls in Hawaii*. Honolulu.
Fisher, Harvey I. 1966. Airplane-albatross collisions on Midway Atoll. *Condor* 68:229–42.
—— and Paul H. Baldwin. 1946. War and the birds of Midway Atoll. *Condor* 48:3–15.
Munro, George C. 1945. Tragedy in birdlife. *'Elepaio* 5:48–49.

17. Conservation at Sea

Harrison, Craig S. 1985. A marine sanctuary in the Northwestern Hawaiian Islands: An idea whose time has come. *Natural Resources Journal* 25:317–47.

Idyll, C. P. 1973. The anchovy crisis. *Scientific American* 228:22–29.

Ohlendorf, Harry M., and Craig S. Harrison. 1986. Mercury, selenium, cadmium and organochlorines in eggs of three Hawaiian seabird species. *Environmental Pollution* (ser. B) 11: 169–91.

Ripley, S. Dillon. 1975. The view from the castle. *Smithsonian* 6, no. 1:6.

Sievert, Paul R., Louis Sileo, and Stewart I. Fefer. In press. Prevalence and characteristics of plastic ingested by Hawaiian seabirds. In *Proceedings of the Second International Conference on Marine Debris*, ed. Richard S. Shomura. Honolulu.

U.S. Congress, Senate, Committee on Commerce, Science, and Transportation. 1986. *Pelagic Driftnet Fisheries Issues: Hearings*. 99th Cong., 1st sess.

18. Conservation Dilemmas

Cooper, George, and Gavan Daws. 1986. *Land and Power in Hawaii: The Democratic Years*. Honolulu.

National Research Council. 1986. *Ecological Knowledge and Environmental Problem-Solving*. Washington, D.C.

INDEX

Adaptations to marine life, 48–52
Aircraft-bird collisions, 118–19, 192
Alau Island, Maui, 11, 59
Albatross. *See* Black-footed albatross; Laysan albatross; Short-tailed albatross; Waved albatross
Allen, Joseph, 34
Alteration of habitat, 192
American Association for the Advancement of Science, 224
American Fishery Society, 221
American golden plover, 47
Anchovies, 26, 91, 158, 165, 188, 208
Anous, 176
Ants, 187
Archaeopteryx, 44
Ariyoshi, George, 213
Ascension Island, 68, 89, 91, 154, 162–64
Au, David, 89

Baitfish, 208–9, 218
Baker Island, 145
Balloonfish, 169
Band-rumped storm-petrel. *See* Harcourt's storm-petrel
Barber's Point, Oahu, 12, 132
Barking Sands, Kauai, 13, 119
Bartsch, Paul, 65
Beebe, William, 23, 85
Big-eyed scad, 208
Bills, 50–51, 77–78
Bishop Estate, 219, 226
Black-crowned night heron, 186
Black-footed albatross
  breeding season, 65, 112–13
  breeding success, 69, 79, 81, 112, 116–17
  conservation, 39, 118–19, 208
  courtship, 76, 114–15
  description, 45, 98, 107, 111
  distribution, 46, 109
  feeding, 50, 69, 80, 84–88, 90, 94, 96, 110–12, 116–17
  fledging, 117–18
  growth and development, 79–80, 116–17
  incubation, 79, 115–16
  longevity, 52
  movements, 99–100, 102, 110–11
  nest sites, 70–71, 113–15
  population, 56–57, 61, 64, 110, 118
  pox, 65, 117
  territory, 74–76, 110, 114
  vocalizations, 116
Black noddy
  breeding season, 181–82
  breeding success, 69, 79, 81, 184–86
  conservation, 187–88, 209
  courtship, 77, 183
  description, 51, 176–79
  distribution, 48, 177–78
  feeding, 24, 69, 84, 86, 88–91, 94, 178, 180–81, 185
  fledging, 185
  growth and development, 79–80, 185–86
  incubation, 79, 184
  longevity, 177
  movements, 101–2, 178, 188

Black noddy (*cont.*)
nest sites, 71–73, 78, 181, 184
population, 55–64, 178
territory, 75, 183
vocalizations, 181
Black Point, Oahu, 12, 123
Blue-gray noddy
breeding season, 181–82
breeding success, 79, 81, 184, 186
conservation, 187–88, 193
courtship, 183
description, 176–79
distribution, 48, 177
feeding, 84, 86, 88, 90, 92–95, 181, 185–86
fledging, 185
growth and development, 79–80, 185
incubation, 79, 184
longevity, 177
movements, 101–2, 178
nest sites, 70–72, 182–84
population, 53, 56–57, 62–64, 177
territory, 183
vocalizations, 181
Board of Land and Natural Resources, 202–3, 222, 227
Bombing of colonies, 10, 38, 127, 164, 187, 193
Bonin Islands, Japan, 7, 35–36, 40, 46, 61, 109, 136
Bonin petrel
breeding season, 126, 128–29
breeding success, 64, 79, 81–82, 130–31
conservation, 30, 129, 132–34, 193–94
courtship, 128
description, 51, 122–23, 126, 128
distribution, 30, 46, 123–24
feeding, 50, 69, 79, 86, 88, 90, 92, 94–96, 125, 130–32
fledging, 131–32
growth and development, 79–80, 130–32
incubation, 79, 129–30
longevity, 123
movements, 101, 124
nest sites, 71–74, 78, 126–28
population, 54–57, 62, 124
territory, 128
vocalizations, 128
Booby. *See* Brown booby; Masked booby; Red-footed booby
Breeding, 67–82
synchronization of, 74, 113, 126, 129, 149, 158, 160, 171, 183
Breeding displays, 49
Breeding season, 68–69. *See also under names of individual species*
Breeding success, 80–82. *See also under names of individual species*
Bristlemouths, 25, 94, 125, 181
Bristle-thighed curlew, 47, 193
Brood patches, 78, 115, 129, 149, 160–61, 171, 184
Brown booby
breeding season, 158
breeding success, 52, 66, 78–79, 81, 161, 163
conservation, 163–65
courtship, 77, 159–60
description, 49, 51, 153–54, 156
distribution, 47, 155–56
feeding, 50, 84, 86–90, 94, 157–58, 162–63
fledging, 162–63
growth and development, 79–80, 161–63
incubation, 78–79, 160–61
longevity, 155
movements, 101, 157, 163
nest sites, 70–71, 153, 158–59
population, 54, 56–58, 62–63, 155–56
territory, 159–60
vocalizations, 154
Brown noddy
breeding season, 181–82
breeding success, 79, 81, 184, 186
conservation, 187–88
courtship, 77, 183–84
description, 49, 51, 176–79
distribution, 48, 177–78
feeding, 84, 86, 88–91, 94, 178, 180, 185
fledging, 185
growth and development, 79–80, 185–86
incubation, 79, 184
longevity, 177
movements, 102, 178
nest sites, 70–73, 78, 182
population, 54, 56–58, 62–63, 177
territory, 183
vocalizations, 181
Bulbul, 186
Bulwer, James, 121
Bulweria, 121
Bulwer's petrel
breeding season, 126, 129
breeding success, 79, 81, 130–31
conservation, 30, 132–34, 194
courtship, 77, 128
description, 122–23
distribution, 30, 46, 123–24
feeding, 86, 88, 90, 92, 95, 125, 130–32
fledging, 131–32
growth and development, 79–80, 130–32
incubation, 79, 129–30
longevity, 123
movements, 101
nest sites, 71–73, 126–27
population, 56–62, 124
territory, 64
vocalizations, 77, 128–29

Bureau of Land Management, 196

Cadmium, 93, 188
California current, 7, 21
Cats, 33, 131, 145, 152, 164
Cattle egret, 186
Channel Islands Marine Sanctuary, California, 213
Charadriiformes, 45, 47–48, 175
Chatham Islands, 31
Chinaman's Hat. *See* Mokoli'i Island
Chlorinated hydrocarbons, 134, 188, 209. *See also* DDT/DDE; PCBs
Christmas Island (Pacific), 7, 52, 66, 68, 80–81, 89, 91, 154, 162
Christmas shearwater
- breeding season, 126, 129
- breeding success, 79, 81, 130–31
- conservation, 31, 132–34
- courtship, 128
- description, 51, 122, 126
- distribution, 46, 123
- feeding, 86, 88–90, 95, 124, 130–32
- fledging, 131–32
- growth and development, 79–80, 130–32
- incubation, 79, 129–30
- longevity, 123
- movements, 100
- nest sites, 70–73, 126–27
- population, 56–58, 61–62, 124
- territory, 127–28
- vocalizations, 128

Ciguatera, 66
Citizens' suits, 226
Clean Water Act, 214
Clipperton Island, 155, 164, 194
Clutch size, 78, 160
Coastal Zone Management acts, 203
Coleridge, Samuel Taylor, 107
Colonial breeding, 73–75
Columbus, Christopher, 41, 144, 168
Commercial Pacific Cable Company, 17, 38
Conant, Sheila, 140
Consumption of marine resources, 95–97, 110, 125, 138, 145, 156, 168
Convergences, ocean, 20, 23
Cook, Captain James C., 32–33
Cooper, George, 219, 225
Coral reefs, 7–8, 18
Courtship, 76–77. *See also under names of individual species*
Cowfish, 84, 90, 92–93, 146, 176, 181
Crater Hill, Kauai. *See* Kilauea Point National Wildlife Refuge
Critical habitat, 200–201
Currents, 20–23
Customary international law, 210, 214
Cypselurus, 89, 146, 169

Daito Islands, Japan, 7, 109
Dall, William, 109
Dark-rumped petrel
- breeding season, 126, 129
- breeding success, 79, 81, 130–31
- conservation, 31, 33, 64, 132–34, 196, 202, 208
- courtship, 77, 128
- description, 51, 122
- distribution, 11, 31, 46, 123–24, 133, 201–2
- feeding, 50, 86, 92, 125, 130–32
- fledging, 131–32
- growth and development, 79–80, 130–32
- incubation, 79, 129–30
- movements, 101
- nest sites, 71–72, 126–27
- population, 55–57, 59–62, 124
- territory, 127–28
- vocalizations, 77, 128–29

Darwin, Charles, 220
Darwin point, 8
Daws, Gavan, 219, 225
DDT/DDE, 134, 164, 188, 209–10
Deep-sea mining, 210, 214–15
Deferred maturity, 52
Department of Land and Natural Resources (DLNR), 38–40, 54, 197, 202, 204–6, 211, 215, 218, 221–24
Desertion of young, 131, 140–41
Diamond, A. W., 152
Diets, 84–93. *See also* feeding *under names of individual species*
Dillingham Air Field, Oahu, 12, 61, 119
Diomedeidae, 108
Disease, 33, 65–66, 75, 174, 195. *See also* Malaria, avian; Pox, avian
DLNR. *See* Department of Land and Natural Resources
Dogs, 8, 194
Dole, Sanford B., 178
Dolphinfish, 26, 68, 84, 89–91, 103, 157, 169, 180–81
Douglas, Captain William, 34
Drift gillnets, 119, 134, 208, 215, 217–18
Droopwing, 117
Dry Tortugas, Florida, 187–88

Eastern Island, Midway. *See* Midway Islands
Easy Rider, 135
Eddies, 21–23
Emory, K. P., 32
Emperor Seamounts, 6–8, 29, 44–45
Endangered species, 46, 62, 132, 141, 200–201, 205

Endangered Species Act
federal, 200–201, 216, 218, 226–27
State of Hawaii, 204, 216, 226–27
Enforcement of laws, 205–6, 216, 220
Environmental impact statement, 198–200, 204, 210, 214, 216
Environmental Policy Act
federal, 198–99, 214, 220
State of Hawaii, 204–5
Environmental Protection Agency, 214
Equatorial countercurrent, 7
Estuarine sanctuary, 213
Evermann, Barton, 38
Evolution of seabirds, 44–45, 52
Exclusive economic zone, 211–12, 214, 216–18

Feather hunting. *See* Millinery trade
Federal lands in Hawaii, 196–97
Feeding flocks, 73, 157, 169, 180, 188. *See also* Tuna birds
Feeding methods, 85–92
Fidelity, 76, 127, 148, 159
Filefish, 146
Finch. *See* Laysan finch; Nihoa finch
Fish and Wildlife Conservation Act, 221
Fisher, Harvey, 112
Fisher, Walter K., 38, 114, 148
Fisheries, 39, 96–97, 110, 119, 134, 165, 174, 188, 195, 208–9, 215–18
Fishery Conservation and Management Act, 216–18
Flat Island, Oahu. *See* Popoi'a Island
Flies, 3, 16, 147, 154
Flyingfish, 26–27, 68, 85, 89–93, 96, 104, 124, 146, 156, 180–81
Cuvier's, 157
Linne's, 89, 146, 157, 169
short-winged, 157
Flyingfish eggs, 26, 69, 84, 87–88, 90, 94, 96, 117
Flying gurnard, 181
Flying squid. *See* Squid
Food supply, 68
Fossil seabirds, 31, 44–45, 121, 153, 166, 175
Fregata, 144
French Frigate Shoals, 5–6, 14, 24, 33–40, 56, 69–70, 72, 97, 110, 209
Frigatebird. *See* Great frigatebird; Lesser frigatebird; Magnificent frigatebird
Fronts, 23, 25
Fry, D. Michael, 209

Gadfly petrel, 120–34
Galápagos Islands, 46–47, 68, 85, 87, 89, 108, 112, 136, 154, 157, 162–63
Gambierdiscus toxicus, 66

Gardner Pinnacles, 5, 15, 34, 56
Gillnets. *See* Drift gillnets
Gilmartin, William, 14
Goatfish, 26, 68, 84–85, 89–92, 96, 124, 134, 165, 180–81, 188, 208
Goats, 33, 194
Gobies, 91, 180–81
Gooney bird. *See* Black-footed albatross; Laysan albatross
Gould, Patrick J., 103, 186
Grant, Gilbert, 141
Gray-backed tern
breeding season, 181–82
breeding success, 79, 81, 146, 184, 186
conservation, 187–88
courtship, 77, 183
description, 51, 176
distribution, 48, 177–78, 187
feeding, 84, 86, 88, 90, 92–95, 178, 181, 185–86
fledging, 185
growth and development, 79–80, 185–86
incubation, 79, 184
longevity, 177
movements, 75, 102, 178
nest sites, 70–73, 182
population, 56–58, 62–64, 177
territory, 183
vocalizations, 181
Great frigatebird
breeding season, 52
breeding success, 76, 79, 81–82, 149–51
conservation, 151–52
courtship, 75–77, 144, 147–49
description, 48–49, 51, 143–44
distribution, 47, 145
disturbance by humans, 150
feeding, 50, 81, 86–90, 95, 103, 146–47, 150–51, 186
fledging, 150
growth and development, 79–80, 149–51
incubation, 79, 149
longevity, 145
movements, 101, 145–46
nest sites, 71–72, 78, 147–49, 159
polygamy, 76
population, 54, 56–58, 62, 145, 151
territory, 74–76, 149
vocalizations, 147–48
Gregory's fish, 180
Grigg, Richard W., 8
Grounding of fishing vessels, 134, 195, 209
Group adherence, 74
Growth and development, 79–80. *See also under names of individual species*
Guano, 23, 34–36, 72, 149, 184
Guilds, feeding, 85–93

Guinea pigs, 36
Gulf of Farallon Marine Sanctuary, California, 213
Gulls, 47–48

Habitat, nesting, 63–65, 69–73
Hadden, F. C., 146
Hakuhe'e Point, Maui, 59
Halawa Valley, Molokai, 30
Haleakala National Park, Maui, 126–27, 132, 196, 201
Halfbeaks, 89–90, 146, 156–57, 180
Hanakapi'ai, Kauai, 13, 178
Hanapepe Valley, Kauai, 13, 136, 139, 142
Harcourt's storm-petrel
  attraction to lights, 137
  breeding season, 138–39
  breeding success, 79, 81, 140–41
  conservation, 30, 141–42, 205
  courtship, 139
  description, 45, 136–37
  distribution, 30, 46, 136–37, 202
  feeding, 80, 92, 137–38
  fledging, 141
  growth and development, 79–80, 140–41
  incubation, 79, 140
  longevity, 137
  movements, 101, 138
  nest sites, 72, 139–40
  population, 56–57, 59, 61, 136
  territory, 139
  vocalizations, 77, 139
Harrison, Peter, 175
Hasegawa, Hiroshi, 109
Hatchetfish, 25, 90, 92, 94, 96, 125, 138
Hauola Gulch, Lanai, 12, 30, 168
Hawaii, island of, 6, 10, 33, 60
Hawaiian archipelago, 3–5
Hawaiian Islands Bird Reservation, 37, 189
Hawaiian Islands National Wildlife Refuge, 196–97, 213
Hawaiian petrel. *See* Dark-rumped petrel
Hawaiians. *See* Polynesians
Hawaiian stilt, 47
Hawaii Volcanoes National Park, 133
Hawkfish, 181
Heat stress, 70, 72, 78, 117, 127, 147, 151, 162, 170, 184
Henderson Island, 31
Herring, 91, 180–81, 188, 208
Hesperornis, 44
Hida, Thomas S., 84
Hidejima (Hide Island), Japan, 46, 136
High flight, 77, 183
Hippoboscid flies, 147, 154
Homing, 113, 127
Honeymoon period, 77, 129, 149
Hono O Na Pali Natural Area Reserve, Kauai, 13, 133, 202
Huelo Island, Molokai, 11, 60
Hulu Island, Maui, 11, 59
Humboldt current, 20, 23
Hump-backed whale marine sanctuary, 213
Hybridization, 76, 108, 154

Ichthyornis, 44
'Ilio Point, Molokai, 11, 178
Imber, Michael, 125
Incubation, 78–79. *See also under names of individual species*
Infertility, 79, 81, 130, 173, 184
Intergovernmental Maritime Organization, 213
International Council for Bird Preservation, 208
International Longshoremen's and Warehousemen's Union, 222
International Whaling Commission, 218
Ironwood trees, 62, 64, 118, 182, 188
Iwo Islands, Japan. *See* Volcano Islands
Izu Islands, Japan, 36, 46, 109, 136

Jacks, 91, 180
Japan, 134, 199, 208–9, 215, 217–18
Jarvis Island, 194
Johnston Atoll, 101, 145, 197
Jordan, David Starr, 38
Jurisdiction, legal, 210–11

Kaemi, Maui, 11, 59
Ka'ena Point Natural Area Reserve, Oahu, 12, 202
Kahinaakalani, Molokai, 11, 178
Kaholo Pali, Lanai, 12, 101, 168, 170, 178
Kahoolawe, 5, 30, 38, 60
Kaluahonu Preserve, Kauai, 196
Kaluapuhi Pond, Oahu, 180
Kanaha Pond, Maui, 11, 180
Kanaha Rock, Molokai, 11, 60
Kaneohe Marine Corps Air Station, Oahu, 61, 70, 119, 164, 193, 197
Kaohikaipu Island, Oahu, 12, 58, 178
Kapapa Island, Oahu, 12, 58
Kapiolani Park, Oahu, 9, 178
Kauai, 6, 13, 57, 61, 109
Kaula Island, 9, 31, 38, 57, 104, 164, 187, 193
Kaunolu Bay, Lanai, 12, 178
Kawaihoa Point, Niihau, 155
Kawainui Marsh, Oahu, 43, 225
Kawakiu Niu Bay, Molokai, 11, 119
Kealakekua Bay, Hawaii, 30
Keaoi Island, Hawaii, 10, 61
Kekepa Island, Oahu, 12, 58
Keopuka Rock, Maui, 11, 59

Key Largo National Marine Sanctuary, Florida, 213
Ki'ei Island, Lanai, 12, 60, 155
Kihewamoku Island, Oahu, 12, 58
Kilauea Crater, Hawaii, 133
Kilauea Point National Wildlife Refuge, Kauai, 13, 70, 145, 159, 164, 168, 170, 196, 218, 226
King, James A., 34
Kleptoparasitism, 86, 95, 146–47, 160–61
Kohala Mountains, Hawaii, 10, 133
Koko Guyot, 6–7
Koko Head, Oahu, 9, 178
Koolau Mountains, Oahu, 6, 12
Korea, Republic of, 134, 208, 217–18
Kotzebue, Otto von, 109
Kridler, Eugene, 195
Krill, 65, 93
K-strategy, 78
Kukuihoolua, Oahu, 58
Kumoa Gulch, Lanai, 12, 60, 133
Kure Atoll, 5, 8, 26, 34–39, 57, 72, 75, 134, 202
Kuroshio, 7, 21, 94, 165

Lanai, 12, 60
Land and Power in Hawaii (Cooper and Daws), 219
Land Use Commission, 203, 222
Land use planning, 202–4
Lanternfish, 25–26, 90, 92, 94, 96, 125
La Perouse Pinnacle. *See* French Frigate Shoals
Laridae, 175
Laupahoehoe Park, Hawaii, 10, 61, 178
Law of the Sea Treaty, 210–11, 213–14, 217
Laws
  enforcement of, 205–6, 216, 220
  federal, 195–201
  State of Hawaii, 201–5
Laysan albatross
  breeding season, 52, 112–13
  breeding success, 66, 79–81, 112, 116–17
  conservation, 39, 118–19, 208–9
  courtship, 67, 76, 114–15
  description, 51, 107–8, 111
  distribution, 46, 61, 109, 202
  feeding, 50, 69, 80, 86–88, 90, 94, 96, 110–12, 116–17
  fledging, 117–18
  growth and development, 79–80, 116–17
  incubation, 79, 115–16
  longevity, 52, 109
  movements, 99–100, 102, 110–11
  nest sites, 70–71, 113–15
  population, 54, 56–58, 60–61, 64, 104, 110, 118
  pox, 65, 117
  territory, 75–76, 110
  vocalizations, 67, 116
Laysan duck, 16
Laysan finch, 16–18, 81, 131, 187
Laysan Island, 3, 5, 15–16, 34–39, 56–57, 70, 72, 101, 110, 118, 134, 195
Laysan rail, 18, 39
Lehua Island, 5, 9, 57, 145
Lesser frigatebird, 47, 144–45
Lights, attraction of birds to, 131, 133, 142, 192
Lindblad Explorer, 152
Lisianski, Captain Urey, 1, 34
Lisianski Island, 1, 5, 15, 34–39, 56–57, 70, 72, 118, 127, 134, 209
Lizardfish, 26, 85, 90–92, 96, 180–81
Loihi Seamount, Hawaii, 6
Longevity, 52. *See also under names of individual species*
LORAN station, 18, 38–39, 74, 157, 188

MacArthur Foundation, 226
Mackerel scad, 26, 68, 85, 89–91, 93, 96, 104, 124, 134, 146, 156–57, 165, 169, 174, 180, 188, 208–9
Magnificent frigatebird, 144, 152
Mahimahi. *See* Dolphinfish
Makaopuhi Crater, Hawaii, 10, 133
Makaweli Beach, Kauai, 13, 136
Malaria, avian, 18, 33, 39, 66, 132, 195, 224
Manana Island, Oahu, 9, 12, 31, 33, 38, 58, 61–62, 69, 75, 159, 173, 188, 202, 225
Manganese crusts, 210
Maniania Pali, Ka'u, Hawaii, 61
Mansfield, George, 35
Maoris, 31
Marcos, Ferdinand, 219
Marcus Island, Japan, 7, 36, 113, 118
Marijuana, 219
Marine debris, 87, 104, 112, 156, 207, 209, 215. *See also* Plastic
Marine life conservation district, 211
Marine sanctuary, 213, 218
Marioris, 31
Maro Reef, 5, 15, 34
Marquesa Islands, 4
Masked booby
  breeding season, 158
  breeding success, 52, 66, 76, 78–79, 81, 161, 163
  conservation, 163–65, 208–9
  courtship, 77, 159–60
  description, 51, 153–54, 156
  distribution, 47, 155
  feeding, 50, 86–90, 94–95, 157–58, 162–63
  fledging, 162–63
  growth and development, 79–80, 161–63
  incubation, 78–79, 160–61
  longevity, 155
  movements, 101, 156, 163

nest sites, 70–71, 74, 78, 158
population, 54, 56–58, 62, 155–56
territory, 74, 76, 159–60
vocalizations, 154
Maui, 11, 33, 59
Maui Nui, 6
Mauna Kea, Hawaii, 31, 132–33
Maunalei Gulch, Lanai, 12, 168
Mauna Loa, Hawaii, 31, 133
Meiji Guyot, 6–7, 45
Mercury, 134, 165, 188, 209–10
Mice, 18
Middens, 31–32, 132, 140–41
Midway Islands, 5, 17–18, 34–40, 57, 61, 69–70, 72, 75, 110, 118–19, 134, 163, 197
Migration, 48, 64, 99–102, 124
Migratory Bird Treaty Act, 199–200, 216, 218, 220
Miles, Dorothy, 182
Millinery trade, 36–37, 56, 62, 66, 118, 173, 187
Minami-Kojima, Senkaku Retto, Japan, 109, 111
Minami Torishima, Japan. *See* Marcus Island
Mokapu Island, Molokai, 11, 60
Moke'ehia Island, Maui, 11, 59
Mokolea Rock, Oahu, 12, 58
Mokoli'i Island, Oahu, 9, 12, 58, 70, 168, 195
Moku 'Ae'ae, Kauai, 13, 57, 119, 145, 155
Mokualai, Oahu, 58
Moku'auia Island, Oahu, 9, 12, 58, 195
Moku Hala, Maui, 11, 59
Mokuho'oniki, Molokai, 11, 60, 202
Moku Lua Islands, Oahu, 12, 58
Moku Mana, Maui, 11, 59
Moku Manu, Oahu, 9, 12, 58, 69, 75, 109, 119, 145, 155, 202
Moku Manu, Molokai, 11, 60
Moku Naio, Lanai, 12, 60
Moku Puku Island, Hawaii, 10, 61
Molokai, 11, 30, 33, 60–61, 109
Molokai Pali, 11, 133
Molokini Island, Maui, 11, 59
Mongooses, 33, 131, 133, 164, 194–95
Monk seals, 11, 13–18, 35, 135
Mosquitos, 18, 33, 39, 119, 132, 195
Munro, George C., 17, 30, 35, 37, 105, 119, 136, 155, 194, 220
Murphy, Robert C., 143, 158
Mutton birds, 31, 132, 191
Mynah, common, 131, 186

Nanahoa Island, Lanai, 12, 60
Na Pali coast, Kauai, 13, 77, 101, 136, 155, 168
National Audubon Society, 226
National Environmental Policy Act. *See* Environmental Policy Act
National Marine Fisheries Service, 38, 84, 198, 201, 224
National Ocean and Atmospheric Administration, 213–15
National Park Service, 196, 201
Natural area reserve system, 202, 211
Natural Resources Defense Council, 226
Nature Conservancy, 196, 225–26
Necker Island, 5, 13–14, 31–34, 56, 72–73
Needlefish, 153, 157, 180
Nelson, J. Bryan, 147–48
Newell's shearwater
attraction to lights, 131, 133
breeding season, 79, 126, 129
breeding success, 81, 130–31
conservation, 30, 64, 132–34, 196, 200–201
courtship, 77, 121, 128
description, 122, 126
distribution, 13, 30, 46, 123–24, 133, 202
feeding, 86, 89, 124, 130–32
fledging, 131–32
growth and development, 79–80, 130–32
incubation, 79, 129–30
longevity, 123
movements, 100
nest sites, 71–72, 126–27
population, 55, 57–61, 124
pox, 133, 195
territory, 127
vocalizations, 77, 121, 128–29
Nihoa finch, 13, 81, 131, 187
Nihoa Island, 5, 11–13, 31–34, 56, 70, 72–73, 134, 141, 193
Nihoa millerbird, 13
Niihau, 5, 30, 57, 109, 119
Niño, El, 66, 80–81, 104, 112, 208
Nitrates in seawater, 23–24
Noddy. *See* Black noddy; Blue-gray noddy; Brown noddy
North equatorial current, 7, 21
North Pacific current, 7, 21
North Pacific subtropical convergence, 7, 20, 101
Northwestern Hawaiian Islands, 4–8, 10–18, 29, 31–40
Nuclear waste disposal, 192
Nutrients, ocean, 19–24, 28

Oahu, 12, 33, 58, 109
Oceanitidae, 135
Oceanodroma, 136
Ocean thermal energy conversion (OTEC), 63, 210, 213–14
Odontopteryx, 45
Ogasawara Islands, Japan. *See* Bonin Islands
Oil glands, 49, 146, 179
Oil pollution, 209, 213
Okala Island, Molokai, 11, 60

Olfactory bulb, 125
Olokele Canyon, Kauai, 13, 136
Olson, Storrs L., 8, 31
Ommastrephes bartrami, 27
Ommastrephidae. *See* Squid
Otata Island, New Zealand, 225
OTEC. *See* Ocean thermal energy conversion
Outer continental shelf, 210–11, 214
Oyashio, 7, 21

Pacific Ocean Biological Survey Program, 38, 99
Pacific plate, 5, 8
Pacific saury, 26, 89–90, 146, 157, 165, 169
Pakalolo, 219
Palila v. *Department of Land and Natural Resources*, 200
Palmer, Henry, 17, 37, 98
Pan American Airlines, 17, 38, 118
Paokalani Island, Hawaii, 10, 61, 178
Papanui O Kane, Maui, 11, 59
Paty, John, 34
Pauuwalu Point, Maui, 59
PCBs, 134, 165, 188, 209–10
Pearl and Hermes Reef, 5, 15–16, 34–39, 57, 70, 72, 101, 112, 193
Pelecaniformes, 45–47
Pelekunu Valley, Maui, 11, 132–33, 168
Penguin Banks, 6
Penguins, 45–46, 48
Pennant, Thomas, 83
Pesticides, 209–10
Petrel. *See* Bonin petrel; Bulwer's petrel; Dark-rumped petrel; Gadfly petrel; *species listed under* Storm-petrel
Phaethon, 167
Phaethontidae, 166
Phosphates, 23
Photosynthesis, 22–24
Phytoplankton, 21–25
Pigs, 8, 30, 33, 119, 164
Pioneer Bank, 16
Piracy, 86, 95, 146–47, 160–61
Pitman, Robert, 89
Plastic, 26, 87, 104, 112, 134, 142, 174, 207, 209, 215
Plate tectonics, 4
Pollution, 209–10
Polygamy, 76
Polynesians, 8, 18, 30–33, 91, 107, 132, 187, 194
Po'opo'o Island, Lanai, 12, 60
Popoi'a Island, Oahu, 9, 12, 58, 195
Populations, 53–66, 208. *See also* population *under names of individual species*
  regulation of, 63–65
  variation in, 65–66
Porpoises and seabirds, 26, 89, 91, 100
Pox, avian, 65, 75, 117, 119, 133, 173, 195
Predators, 8, 33, 64, 74, 81, 127, 130–34, 141, 151, 163–64, 173, 186–87, 194, 220. *See also* Cats; Dogs; Mongooses; Pigs; Rats
Prelaying exodus, 129, 140
Primary productivity, 21–25, 52, 68, 81, 97
Procellariidae, 121
Procellariiformes, 46
Protoavis, 44
Pterodactyl, 43–44
Pterodroma, 121
Puffinus, 121
Pulemoku, Oahu, 58
Pumice, 209
Pu'u Ha'ao, Hawaii, 10, 133
Pu'u Kanakaleonui, Hawaii, 10, 133
Pu'u Koa'e, Kahoolawe, 60
Pu'uku, Maui, 11, 59
Pu'u Pehe, Lanai, 12, 60
Pu'u 'U'au, Oahu, 12, 132

Rabbit Island, Oahu. *See* Manana Island
Rabbits, 36, 38, 62, 118, 134, 151, 164, 173, 194
Rats, 129, 187, 194–95, 206, 224
  black, 18, 33, 39, 61–62, 81, 133, 141, 173, 187, 194
  Norwegian, 33, 194
  Polynesian, 8, 18, 30, 81, 119, 131, 133, 141, 152, 173, 187, 194
Rauzon, Mark J., 140–41, 194
Reagan, Ronald W., 211
Recolonization, 61–62, 74–75, 109, 113, 118–19, 145, 164, 167, 178, 188
Red-billed tropicbird, 47
Red-footed booby
  breeding season, 158
  breeding success, 66, 78–79, 81, 161, 163–64
  conservation, 31, 163–65, 193, 209
  courtship, 77, 159–60
  description, 51, 153–56
  distribution, 47, 155, 164
  feeding, 86–90, 94, 157–58, 162–63
  fledging, 162–63
  growth and development, 79–80, 161–63
  incubation, 78–79, 160–61
  longevity, 155
  movements, 101, 156–57, 163
  nest sites, 71–73, 78, 159
  population, 54–58, 62–63, 155–56
  territory, 159
  vocalizations, 154–55
Red-tailed tropicbird
  breeding season, 169–70
  breeding success, 79, 81, 103, 172–73
  conservation, 173–74

courtship, 76–77, 166, 171
description, 49, 51, 166–68
distribution, 47, 167–68
feeding, 80, 86–90, 103, 168–69, 172
fledging, 172
growth and development, 79–80, 172
incubation, 79, 171–72
longevity, 167
movements, 101, 168–69
nest sites, 70–73, 78, 170–71
population, 54, 56–60, 62, 64, 167–68
pox, 65, 173
territory, 75, 170, 173
vocalizations, 167, 169, 171–72
Research natural area, 39, 197
Rest years, 68, 158
Revilla Gigedo Islands, 109
Rhodopsin, 94, 111, 124–25
Ripley, S. Dillon, 218
Romer, Alfred S., 44
Roosevelt, Theodore, 37
Rothschild, Walter, 17, 37, 39, 98, 128
R-strategy, 78
Rudderfish, 157
Ruddy turnstone, 47
Ryukyu Islands, Japan, 7, 109

St. Matthew Island, Alaska, 197
Salinity, ocean, 19, 21–22
Salmon Bank, 5
Salt glands, 50
Sanderling, 47
Sand Island, Midway. *See* Midway Islands
Sand Island, Oahu, 155–56
Schauinsland, H. H., 37
Schlemmer, Max, 36
Schreiber, Ralph W., 152
Seabird sanctuary, State of Hawaii, 39–40, 58, 202
Sea Life Park, Oahu, 155, 164
Seal-Kittery Island. *See* Pearl and Hermes Reef
Sea striders, 92–93, 138, 181
Seki, Michael P., 84
Senkaku Retto, Japan, 109
Seychelle Islands, 68, 89, 91, 154, 162–63, 187
Shark, 24, 81–82, 103, 117–18
Shearwater. *See* Christmas shearwater; Newell's shearwater; Short-tailed shearwater; Wedge-tailed shearwater
Short-tailed albatross, 46, 75–76, 87, 108–10, 118–19, 200–201, 208
Short-tailed shearwater, 124
Sibling murder, 78, 153, 160
Sierra Club Legal Defense Fund, 226
Sievert, Paul R., 134, 209
Sikes Act Extension, 197
Simons, Theodore, 129
Sincock, John L., 129, 136
Site fidelity, 74, 76
Snake mackerel, 96, 180
Sooty storm-petrel
attraction to lights, 135, 142
breeding season, 138–39
breeding success, 64, 79, 81–82, 139–41
conservation, 141–42, 200
courtship, 139
description, 51, 135–37
distribution, 46, 136
feeding, 50, 69, 80, 86, 88, 90, 92, 137–38
fledging, 141
growth and development, 79–80, 140–41
incubation, 79, 140
longevity, 137
movements, 101, 138
nest sites, 71–73, 139–40
population, 56–57, 137
territory, 139
vocalizations, 77, 139
Sooty tern
breeding season, 181–82
breeding success, 69, 79–81, 103, 146, 184, 186
conservation, 187–88, 192, 209
courtship, 77, 183
description, 49, 175–77, 179–81
disease, 66
distribution, 48, 177–78
feeding, 79–80, 86–91, 94–96, 103, 177–80, 185–86
fledging, 185–86
growth and development, 79–80, 185–86
incubation, 79, 184
longevity, 52, 177
movements, 74–75, 102, 178, 188
nest sites, 70–71, 73, 78, 182
population, 54–58, 62–63, 177
territory, 183, 186
vocalizations, 77, 176, 181, 185
Southeast Island. *See* Pearl and Hermes Reef
Southern oscillation, 66
Sphenisciformes, 45
Squid, 25–27, 85, 87–89, 91–97, 124, 146, 157, 165, 169, 174, 188, 209, 218
Squirrelfish, 91, 124
Star fracture, 116, 130, 172, 185
Starvation, 81, 117, 140, 163, 174, 186
State lands, 202
State plan, Hawaii, 202, 221
Steller, George, 108
Steller's sea eagle, 74
Sterninae, 175
Sthenoteuthis oualaniensis, 27
Stomach oil, 80, 116, 130, 140, 142
Stomatopods, 169, 181

Storm-petrel. *See* Harcourt's storm-petrel; Sooty storm-petrel
Storms, 66, 68–69, 81, 102–4, 112, 117, 131, 158, 182, 184, 186
Styrofoam, 87, 209
Submerged Lands Act, 210
Sula, 153–54
Sulidae, 153
Sunfish, 89–90, 169
Survival at sea, 102–4
Swordfish, 26
Syrinx, 154

Taiwan, 134, 208, 217–18
Tanager expedition, 38
Temperature, ocean, 19–24
Temperature regulation, 79, 151
Tern. *See* Gray-backed tern; Sooty tern; White tern
Tern Island. *See* French Frigate Shoals
Territorial sea, 211
Territory, 75–76. *See also under names of individual species*
Thermocline, 21–24
Threatened species, 46, 200
Ticks, 49, 121
Tinker, Spencer, 180
Tocqueville, Alexis de, 225
Torishima Island, Japan, 75, 109–10, 141
Tourists, 193
Townsend Cromwell, 3, 98, 120
Townsend's shearwater. *See* Newell's shearwater
Tripartite Research Program, 38
Trophic level, 24–25
Tropical convergence, 7, 20
Tropicbird. *See* Red-billed tropicbird; Red-tailed tropicbird; White-tailed tropicbird
Truman, Harry S., 39
Trust for Public Land, 226
Tuna, 24–27, 68, 89–92, 103, 134
  little, 27, 91, 180
  skipjack, 26–27, 89, 91, 208, 216
  yellowfin, 26–27, 89, 91, 100, 208
Tuna birds, 85, 89, 91
Turtles, 13–18, 35, 104, 127, 146, 156, 163, 173, 179

Uluhe ferns, 72, 126, 133
Ulupa'u Head, Oahu, 12, 155, 159, 164, 193
Upwelling, 20, 23, 25
U.S. Coast Guard, 18, 38–39, 74, 192, 215, 217
U.S. Fish and Wildlife Service, 38–40, 54, 110, 141, 196, 198–99, 201, 207, 218, 221–24
U.S. Forest Service, 196
U.S. Navy, 10, 40, 187, 193, 197, 215
USSR, 199

Vertical migration, 25, 94
Volcano (Iwo) Islands, Japan, 46, 118, 136
Volcanoes, 4–8, 109

Waianae Mountains, Oahu, 6, 12
Waihe'e Point, Maui, 9, 11, 59
Waimanu Valley, Hawaii, 10, 133, 213
Waimea Canyon, Kauai, 13, 77, 101, 168
Waipio Valley, Hawaii, 10
Wake Island, 7, 36, 101, 113, 118, 156
Wandering tattler, 47
Watson, John B., 183
Watt, James, 197
Waved albatross, 87, 108
Weather. *See* Storms
Wedge-tailed shearwater
  breeding season, 126, 129
  breeding success, 69, 76, 79, 81, 130–31
  conservation, 30, 132–34, 193
  courtship, 77, 120–21, 128
  description, 51, 99, 121–22, 126
  distribution, 13, 30, 46, 123
  feeding, 84, 86, 88–91, 94–96, 124, 130–32
  fledging, 131–32
  growth and development, 79–80, 130–32
  incubation, 79, 129–30
  longevity, 123
  movements, 99–100, 102, 124
  nest sites, 71–74, 126–28
  population, 54–61, 124
  territory, 64, 73, 75
  vocalizations, 77, 120–21, 128–29
Weights of seabirds, 52. *See also* description *under names of individual species*
Wellwood, M/V, 213
Western Pacific Fisheries Management Council, 216–19
Wetmore, Alexander, 18
Whales and seabirds, 89, 112, 125
White-tailed tropicbird
  breeding season, 169–70
  breeding success, 79, 81, 172–73
  conservation, 173–74
  courtship, 77, 171
  description, 166–68
  distribution, 47, 167–68, 202
  feeding, 50, 86–87, 95, 168–69, 172
  fledging, 172
  growth and development, 79–80, 172
  incubation, 79, 171–72
  longevity, 167
  movements, 101, 169
  nest sites, 70–71, 170
  population, 57–61, 167–68
  territory, 170, 173
  vocalizations, 167, 169, 171–72

White tern
  breeding season, 181–82
  breeding success, 79, 81, 184, 186
  conservation, 187–88, 205, 209
  courtship, 77, 183
  description, 49, 51, 175–76, 178–79
  distribution, 9, 48, 177–78
  feeding, 79–80, 84, 86, 88–91, 94, 178, 180, 185
  fledging, 185
  growth and development, 79–80, 185–86
  incubation, 79, 184
  longevity, 177
  movements, 102, 178, 188
  nest sites, 71–73, 78, 183–84
  population, 53, 56–58, 62, 64, 177
  territory, 74–75, 183
  vocalizations, 175, 181
Whittow, G. Causey, 79
Wildlife cooperative research unit, 221
Wildlife Management Institute, 221
Wildlife plan, State of Hawaii, 205–6
Wildlife refuge, 39–40, 196–97, 199
Wind sailors, 96
Wires, bird collisions with, 39, 188, 192
Woodside, David, 174, 194
World War II, 14, 38–39, 61, 107, 114, 118, 163, 187, 193

Yamashina, Yoshimaro, 109

Ziegler, Alan C., 132, 139